FILE COPY
FDK.

Curtis Brown Ltd
1 Craven Hill
London
W2 3EW

ISBN 0 7153 6407 3

Library of Congress Catalog Card Number 74-83558

© K. Holden and A. W. Pearson 1975

Set in 11pt on 12pt Imprint (Series 101)
and printed in Great Britain
by Redwood Burn Limited, Trowbridge & Esher
for David & Charles (Holdings) Limited
South Devon House Newton Abbot Devon

Published in the United States of America
by David & Charles Inc
North Pomfret Vermont 05053 USA

Published in Canada
by Douglas David & Charles Limited
132 Philip Avenue North Vancouver BC

Contents

Preface

This book attempts to introduce to students with a very limited mathematical background the essential mathematics needed for a study of economics. It is not intended to replace the more formal mathematics texts and does not include proofs of all the formulae used. These are only included where they can easily be followed. It is hoped that this approach will be more suitable for those students who, as a result of their earlier experiences in the subject area, do not regard themselves as having any mathematical ability. The inclusion of worked examples in the text and exercises with answers in worked form (at the end of the book) are intended to help such students.

The material is selected so as to increase in difficulty as the book progresses. The introductory chapter on Linear Equations leads to the more general chapters on Elementary Matrix Algebra and Non-linear Equations. Chapter 4 on Series deals with applications of immediate relevance as well as providing the groundwork for the chapters on calculus. Chapters 5 and 6 are concerned with the relationship between two variables whilst more general relationships are covered in Chapter 7. This chapter ends with the problem of maximisation subject to constraint, and a variation on the same problem is presented in Chapter 8. The final chapters deal with dynamic relationships, in continuous terms in Chapter 9 and discrete terms in Chapter 10. Trigonometric functions are considered in detail in Appendix A.

The order in which the material is covered can be varied. For example some teachers may prefer to leave Matrix Algebra to the end of a course, or to follow Integral Calculus with Differential Equations.

While the text is intended primarily for students of economics it has proved to be extremely helpful in the teaching of mathematics to students of business studies. It should also be useful to managers who need to renew their acquaintance with the basic mathematical techniques relevant to operational research.

We are grateful to Roger Latham and David Peel of the University of Liverpool for reading through an earlier draft and making many useful suggestions, and also to the copy-editor, Alan Whittle, for his constructive comments on the final draft. They are, of course, in no way responsible for any errors. We are also grateful to Miss Pauline Conley who typed a difficult manuscript in record time.

K.H.
A.W.P.

CHAPTER ONE

Linear Equations

1.1. INTRODUCTION

An equation can be a very simple means of summarising the important features of a particular situation. It may be used for descriptive purposes or it may provide information which is very useful for decision making.

To illustrate this it is useful to look at a particular case. If we choose one from the production-management area we may have a situation in which the total cost of production is determined by the level at which a particular process is operated, ie upon the level of output. This could be expressed as follows:

$$TC = f(q)$$

which is read 'TC is a function of q', where TC represents total cost and q represents quantity or level of output.

The symbol f coupled with the brackets is a shorthand way of saying that the two quantities or *variables* are related in some way, which is as yet unspecified but is assumed to be single valued, ie for each value of q there is only one value of TC.

This may not appear to be very useful and indeed it is not unless we go further and attempt to identify the form which the relationship between the two variables will take. To do this we make two assumptions, which would have to be verified in practice, but which would not be considered unreasonable in many situations. These are as follows:

(a) There are parts of the total cost of production which will not be affected by changes in the level of output because they must necessarily be incurred if the process is adopted and they do not increase as production is increased. These are known as *fixed costs* and include such items as rent, rates, and sections of the labour force which within certain limits are not affected by changes in the level of output.

(b) There are other parts of the total cost of production which increase as the level of output increases. These are known as *variable costs* and include such items as raw materials, power, and parts of the labour force which can be employed on the process if and when required.

It is clear that in many practical situations the split between fixed and variable costs cannot be made very clearly and that in the long run all costs tend to be variable. However, a simple breakdown into these two categories can prove to be very useful in establishing such points as the breakeven level of production, as we shall see later. But first, let us consider an example in which the available information about the production process indicates that the fixed costs amount to £100 and that the variable costs associated with manufacturing 100 units of the product amount to £300. We will assume that the variable costs are directly proportional to the number of units of output and hence that the total cost varies linearly with the level of output.

The relationship between costs and output can then be written

Total cost = fixed cost + (variable cost) × (level of output)

or $$TC = a + b \times q$$

where a and b are two constants which represent fixed cost and variable cost respectively.

1.2 GRAPHICAL REPRESENTATION

For a two-variable relationship a common method of presentation of information is by means of a *graph*. This is a two-dimensional diagram with two *axes*, one of which represents TC (in our example the total cost of production) and the other q (the level of output). These axes are generally drawn at right angles to each other and their point of intersection, the origin, O, is where both total costs and the level of output are zero. By choosing suitable scales to represent different values of the variables we can construct a graph from any given set of data.

In our example when the level of output is 0, the fixed costs and hence the total costs are equal to £100, ie when $q=0$, $TC=100$. Also, when the level of output is 100, the total costs are £400, ie when $q=100$, $TC=400$.

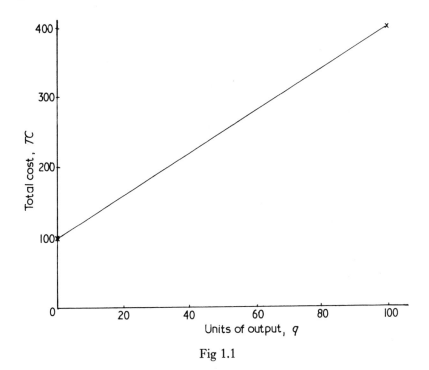

Fig 1.1

Since we are assuming that total cost increases uniformly and continuously with output we can join the two points by a straight line. Hence $TC=a+bq$ is an equation of a straight line, and in this case

$a=$ fixed costs $=£100$ and

$b=$ variable cost per unit of output $=£3$

Effectively we are determining the values of a and b by using the information about the two points to solve the equation $TC=a+bq$.

15

This information enables us to locate two points on the graph with the *coordinates* (0, 100) and (100, 400), where the two numbers in brackets refer to the values of the variables measured along the horizontal and the vertical axes respectively. In general, any point on this graph (Fig. 1.1) is represented by the coordinates (q, TC).

For the first point: $q = 0$, $TC = 100$ and so

$$100 = a + b \times 0$$

$$a = 100$$

and for the second point $q = 100$, $TC = 400$ and so

$$400 = a + 100b$$

Substituting the value $a = 100$ in this equation gives

$$400 = 100 + 100b$$

Subtracting 100 from each side of this equation leaves us with

$$300 = 100b \qquad \text{and so} \qquad b = 3$$

Hence $TC = 100 + 3q$ is the equation which summarises the available information about the production process and describes how costs vary with output.

This equation can now be used to determine the total cost of production at other levels of output. For example, when $q = 20$, $TC = 100 + 3 \times 20 = £160$.

If this point is included on the graph we find that it lies on the straight line joining the two points corresponding to the initial data. We would also find that all other points which satisfy the equation lie on the same straight line. For this reason $TC = 100 + 3q$ is known as a *linear equation*.

In general the choice of symbols for the variables in an equation is made by the people concerned with the presentation and analysis of the information. Some letters tend to be used more frequently than others, and some have fairly agreed usage for particular variables. However, there are no exact rules to be followed, and it is important that the principle is understood, and that emphasis is not placed on the symbols themselves. It follows that $Y = a + bX$ could have been used in the example we have just considered, with Y replacing TC and

16

X replacing q. The form of the equation is unaltered. In the following sections a number of different letters will be used to represent the variables in an equation. This is not intended to confuse, but to assist in understanding that the form of an equation does not depend upon the symbols which represent the variables. However, in all practical cases a clear indication should be given of what the symbols do represent.

It is possible to determine the form of a linear equation from two pairs of values of the variables. This follows graphically because only one straight line can be drawn through any two given points and algebraically because there are two constants to be determined and two pairs of values provide two equations in a and b which can, in general, be solved.

1.3 INTERCEPT AND SLOPE

In the equation $Y = a + bX$, a is often referred to as the *intercept term* because it is the value at which the straight line intercepts the vertical axis, ie it is the value of the function when $X = 0$, and b is often referred to as the *slope* or gradient because it is a measure of the inclination of the line to the horizontal axis.

We can now make use of such an equation in a number of ways. For example we can examine the effect which would be produced by an alteration in a and b, eg in fixed and variable costs, as follows:

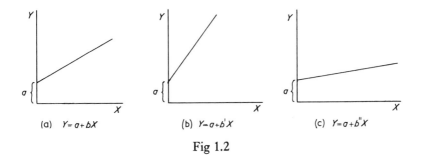

(a) $Y = a + bX$ (b) $Y = a + b'X$ (c) $Y = a + b''X$

Fig 1.2

1. Fixed costs remain the same but variable cost changes. In Fig. 1.2 a is the same in all three cases, but in graph (b) the variable cost is

17

greater than in graph (a) and in graph (c) variable cost is less than graph (a).

That is, b' is greater than b (written $b' > b$)

and b'' is less than b (written $b'' < b$)

The line will then rotate about the point at which it crosses the cost axis.

2. Variable cost remains the same but the fixed costs change. The line will then move up or down parallel to itself, ie with the same slope. In Fig. 1.3 b is constant in all three cases, but graph (b) shows higher fixed costs than graph (a) and graph (c) shows lower fixed costs than graph (a). That is,

$$a' > a \quad \text{and} \quad a'' < a.$$

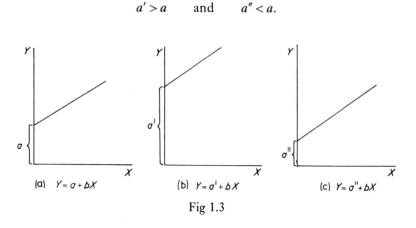

(a) $Y = a + bX$ (b) $Y = a' + bX$ (c) $Y = a'' + bX$

Fig 1.3

1.4 INTERPOLATION AND EXTRAPOLATION

We have seen that a relationship between two variables can be shown graphically, or alternatively it can be represented by a mathematical expression. This equation can then be used to determine the value of total cost at a point in between the points for which we have some empirical data. This is known as *interpolation*.

It is, of course, possible to continue the straight line indefinitely in either direction, ie for negative values of X and for values of X greater than 100. This is known as *extrapolation* and care must be exercised if such a procedure is attempted because the end result may not be a sensible one. For example, values of X which are negative correspond to negative output and have no practical meaning. We must, therefore, eliminate such values by specifying that X must be greater than or equal to zero.

But what about values of X greater than 100? These may or may not be possible, and we have to be careful when we are considering extrapolation that we are certain the fixed and variable costs will remain the same when more than 100 units are produced. If they do not, then we must add the condition that X cannot be greater than 100, or alternatively, X must be less than or equal to 100.

The cost function can then be written

$$Y = 100 + 3X \qquad (0 \leqslant X \leqslant 100)$$

where the constraint, written inside the brackets, implies that the cost function is only a valid representation of the process for outputs from 0 to 100 units.

1.5 EXERCISES

1. Draw the graphs of the following equations for the range $X = 0$ to $X = 6$. What are the intercept term and the slope for each equation?

   (a) $Y = 3X + 4$ (b) $Y = 3X - 4$

   (c) $Y = 4 - 3X$ (d) $Y = 3X$

   (e) $Y = 4$

2. Establish a linear relationship for the total cost function which fits the following two conditions:
   (a) at an output of 10 units the total cost is £70, and
   (b) at an output of 20 units the total cost is £120.
   What do you infer about the fixed and variable cost? Sketch the graph of the total cost function for outputs less than 30.

3. Establish a linear relationship for the total cost function from the following data:
   (a) the variable cost is £6 per unit, and
   (b) the fixed cost is £15.

What output would make the total cost £75? Sketch the graph of the total cost function for outputs less than 20.

4. The total cost of production (Y) of a product is related to the number of units of output (X) by the equation

$$Y = 3 + 2X$$

(a) Sketch the graph of this line for $0 \leqslant X \leqslant 25$
(b) What are the fixed cost and variable cost?
(c) What will be the total cost of producing 15 units of output?
(d) If the total cost is 45, how many units of output will be produced?

1.6 SIMULTANEOUS LINEAR EQUATIONS IN TWO UNKNOWNS

We have seen how to obtain the equation of a straight line from two sets of data. From the data two equations were obtained. These had to be satisfied simultaneously and this enabled the values of a and b to be determined.

In economics we frequently encounter situations in which there are two or more equations which have to be satisfied simultaneously. For illustration let us return to our earlier example in which the cost of production could be represented by a linear equation

$$Y = a + bX$$

This relationship is obviously important but does not tell us anything about the profitability of the overall operation. This can only be assessed if the price is known at which the resulting product might be sold along with the quantity that can be sold at this price.

Under conditions of perfect competition it is usual to assume that the price of the article is fixed and that at this price it is possible to sell all that can be produced. This would be the situation confronting an individual producer whose output formed only a very small part of the total demand. In this case his demand curve would be horizontal (ie price $= p$) (Fig. 1.4).

The revenue function can then be represented by a linear equation which can be written as

Total revenue = price per unit × number of units sold

$$\text{ie } R = pX$$

Thus when $X = 0$, $R = 0$.

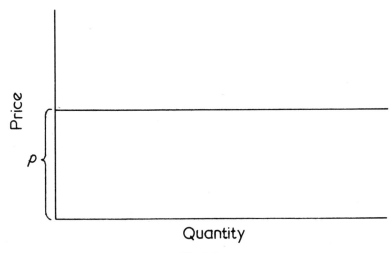

Fig 1.4

Since in this case the revenue function starts at zero and the cost function at the value a, it is obvious that the two lines will only inter-sect at a positive level of output if the constant p is greater than the constant b.

Assuming that this is so we would have the situation in Fig. 1.5.

Notice that we are using one set of axes for two graphs. We can do this because the horizontal axis represents the number of units of output and this scale is the same for both equations. The vertical axis represents cost and revenue respectively for the two equations but these are both measured in £ and, therefore, the same axis with the same scale can be used.

The point K where the lines intersect is often referred to as the *breakeven point* because at this point the total revenue equals the total cost and the net revenue is zero. The volume of output corresponding

21

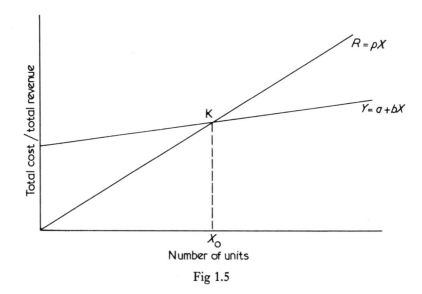

Fig 1.5

to this point is equal to X_0, which is obtained by dropping a perpendicular from the point of intersection to the horizontal axis. If sales are below X_0 a loss will be made and if sales are above X_0 a profit will be made.

The graphs which have been drawn show the cost and revenue equations and from them we can determine the breakeven point.

The same result can also be obtained from the following set of equations:

$$Y = a + bX \qquad (1)$$

$$R = pX \qquad (2)$$

with the added condition that at the breakeven point

$$R = Y \qquad (3)$$

These are three equations in three unknowns (R, Y and X) and are then a complete description of the system. By using eq. (3) we can equate eqs. (1) and (2) to determine X_0, the breakeven point:

$$R = Y$$

$$\therefore \quad pX = a + bX$$

This equation can be solved for X by grouping terms involving X on one side of the equality. Subtracting bX from each side gives

$$pX - bX = a + bX - bX$$

or $\qquad pX - bX = a$

We can now take out the factor X from the terms on the left hand side because it is common to both and obtain

$$X(p - b) = a$$

We then complete the algebraic manipulation by dividing both sides of the equation by the term $(p - b)$ to give

$$X = \frac{a}{(p - b)}$$

None of the changes we have made has in any way altered the relationship between the variables, although we have ended up with an expression which enables us to calculate the breakeven point for the process for any value of p, given that we know the values of a and b. In descriptive terms,

$$\text{Breakeven point} = \frac{\text{fixed cost}}{\text{price} - \text{variable cost}}$$

The difference between the price and the variable cost at a particular level of output is often termed the *contribution* per unit of product, because it is the net revenue which is produced by each extra unit sold. The breakeven point X_0 is equal to the ratio of the fixed cost to the contribution per unit of product. Since the producer's net revenue (represented by N) is the difference between his revenue and his costs, it is given by

$$N = R - Y = pX - a - bX$$
$$= (p - b)X - a$$

Example Total cost of production is given by $Y = 25 + 6X$ and price is fixed at £11. What is the value of output at the breakeven point, and what is the net revenue if 20 units of output are produced?

At the breakeven point

$$\text{total cost} = \text{total revenue}$$

that is,
$$Y = pX = 11X$$

$$\therefore \qquad 25 + 6X = 11X$$

$$25 = 5X$$

$$X = 5$$

At the breakeven point the output is 5 units. The net revenue is given by
$$N = pX - Y = 11X - 6X - 25 = 5X - 25$$

When $X = 20$, $N = 5 \times 20 - 25 = 100 - 25 = 75$. Hence, a net revenue of £75 is obtained when the output is 20 units.

1.7 DEMAND AND SUPPLY

For any particular product there are, in general, a number of potential consumers and the quantity which each consumer buys depends, at least in part, upon the price at which the product is offered on the market. It is safe to assume, therefore, that the quantity demanded of any good is affected by its price and that for most goods the relationship is an inverse one. That is, the quantity demanded decreases as the price is increased and vice versa. This can be expressed mathematically, and in the simplest case it is possible to think of it as being represented by a straight line (Fig. 1.6).

Although the quantity demanded is, by convention, represented along the horizontal axis the equation is usually written algebraically in the form

$$q_d = f(p) = a + bp$$

In this example the constant b would be negative because the line is downward sloping and the relationship is such that the quantity demanded q_d decreases as the price is increased.

In a similar way the market, in a competitive situation, can be considered to be supplied by a large number of independent producers. The number of such producers and the level of their individual outputs for any product is determined by the cost of production

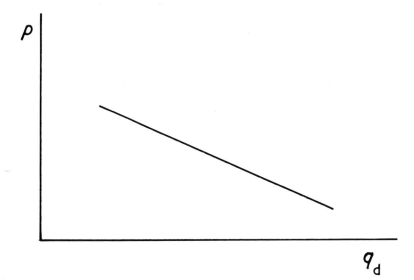

Fig 1.6

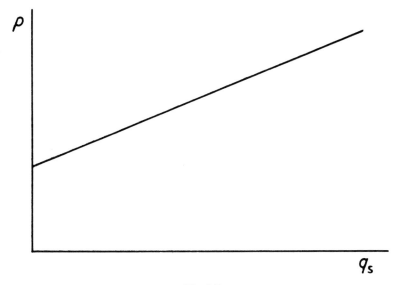

Fig 1.7

25

and the price which the market is prepared to pay for the product. It can be assumed that the supply of any product is affected by the price which can be obtained for it, and, in this case, the quantity supplied increases as the price increases (Fig. 1.7).

Algebraically this can be written

$$q_s = a' + b'p$$

The two equations can be drawn on the one set of axes (Fig 1.8).

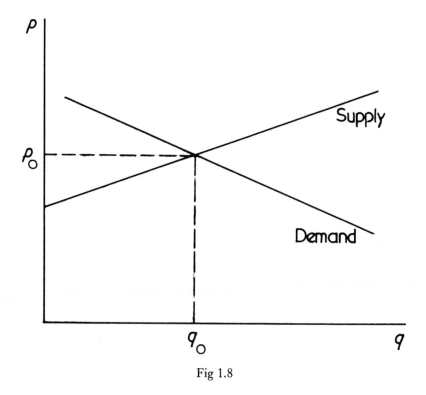

Fig 1.8

The point of intersection of the two lines is the point at which the demand and supply for the product are equal and the market is in equilibrium.

26

This point can be found by solving the pair of simultaneous equations:

$$q_d = a + bp$$

$$q_s = a' + b'p$$

using the information that at the equilibrium point $q_d = q_s$.

From these three equations

$$a + bp_0 = a' + b'p_0$$

Subtracting a and $b'p_0$ from each side and cancelling gives

$$p_0(b - b') = a' - a$$

$$p_0 = \frac{a' - a}{b - b'}$$

and

$$q_d = a + b(p_0)$$

$$= a + b\left[\frac{a' - a}{b - b'}\right]$$

$$= \frac{ab - ab' + ba' - ba}{b - b'} = \frac{ba' - ab'}{b - b'}$$

These examples show one method of solving a system of simultaneous equations. In general, if there are as many equations as unknowns, then it is possible to obtain a unique value for each of the unknowns. Cases do arise, however, when this is not possible, but we will defer discussion of these until Chapter 2.

1.8 EXERCISES

1. The total cost of production (Y) of a product is related to the number of units of output (X) by the equation

$$Y = 33 + 2X$$

(a) What is the breakeven point if the price is fixed at £13 per unit?

(b) What would the net revenue be at an output of 15 units?

(c) Sketch the total cost and revenue functions on the same graph.

(d) Why would the producer not fix the price at £1 per unit?

27

2. A producer has a fixed cost of £50 and a variable cost of £5 per unit of output if his output is less than 200 units.
   (a) What is the equation of his total cost function (which is linear)?
   (b) What is the breakeven point if the producer fixed his price at £10 per unit?
   (c) What is the producer's net revenue if the output is 12 units and the price is (i) £5, (ii) £10, (iii) £15?

3. A producer is willing to supply a market with an output q_s according to the relationship

$$q_s = 25p - 10$$

where p is the price.
The demand function for his product is

$$q_d = 200 - 5p$$

where q_d is the quantity demanded at price p.
   (a) What is the equilibrium price and quantity?
   (b) If the producers supply function changes to

$$q_s = 20p - 25$$

what is the new equilibrium position?

4. Draw the graphs of the following equations and where possible find the values of p and q which satisfy the equations simultaneously.

(a) $2p + 3q = 17$
$5p - 4q = 8$

(b) $6p - q = 3$
$4p + 7q = 2$

(c) $2p + 4q = 7$
$4p + 8q = 14$

(d) $2p + q = 1$
$4p + 2q = 1$

1.9 DETERMINANT NOTATION

We now consider the solution of a general set of simultaneous equations using the methods described above. The general system of two equations in two unknowns is

$$a_{11}x_1 + a_{12}x_2 = b_1 \qquad (1)$$

$$a_{21}x_1 + a_{22}x_2 = b_2 \qquad (2)$$

where the a's and b's are constants. Here, x_1 and x_2 are the unknowns and a_{11} is the coefficient in eq. (1) of x_1 and a_{12} is the coefficient in eq. (1) of x_2 etc. In general, therefore, a_{ij} refers to the coefficient in eq. (i) of x_j.

To obtain the solution for x_1 we eliminate x_2 from the equations by multiplying (1) by a_{22} and (2) by a_{12} to give

$$a_{11}a_{22}x_1 + a_{12}a_{22}x_2 = a_{22}b_1$$

$$a_{12}a_{21}x_1 + a_{12}a_{22}x_2 = a_{12}b_2$$

Subtraction now gives

$$(a_{11}a_{22} - a_{12}a_{21})x_1 = a_{22}b_1 - a_{12}b_2$$

Hence, provided $a_{11}a_{22} - a_{12}a_{21} \neq 0$,

$$x_1 = \frac{a_{22}b_1 - a_{12}b_2}{a_{11}a_{22} - a_{12}a_{21}} \tag{3}$$

and similarly

$$x_2 = \frac{a_{11}b_2 - a_{21}b_1}{a_{11}a_{22} - a_{12}a_{21}} \tag{4}$$

Notice that the denominator in (3) and (4) is the same and that it is a combination of all the a_{ij}-values from (1) and (2). In particular if the pattern of a_{ij}-values from (1) and (2) is reproduced:

$$\begin{matrix} a_{11} & & a_{12} \\ & \times & \\ a_{21} & & a_{22} \end{matrix}$$

we see that the denominator is the cross-product of $a_{11}a_{22}$ minus the cross-product $a_{12}a_{21}$. We now adopt a special notation and define the *determinant* of the numbers a_{ij} as

$$|\mathbf{A}| = \begin{vmatrix} a_{11} & a_{12} \\ a_{21} & a_{22} \end{vmatrix} = a_{11}a_{22} - a_{12}a_{21}$$

It should be noted that the result of multiplying out or evaluating a determinant is a number. For example, if $a_{11} = 4$, $a_{12} = 2$, $a_{21} = 2$, $a_{22} = 3$, then

$$\begin{vmatrix} 4 & 2 \\ 2 & 3 \end{vmatrix} = 4 \times 3 - 2 \times 2 = 12 - 4 = 8$$

29

or if $a_{11} = 2$, $a_{12} = 0$, $a_{21} = 1$, $a_{22} = 3$ then

$$\begin{vmatrix} 2 & 0 \\ 1 & 3 \end{vmatrix} = 2 \times 3 - 0 \times 1 = 6 - 0 = 6$$

From the definition of a determinant it will be clear that

$$\begin{vmatrix} b_1 & a_{12} \\ b_2 & a_{22} \end{vmatrix} = a_{22}b_1 - a_{12}b_2$$

and

$$\begin{vmatrix} a_{11} & b_1 \\ a_{21} & b_2 \end{vmatrix} = a_{11}b_2 - a_{21}b_1$$

and hence (3) and (4) can be written in determinant notations.

$$x_1 = \frac{\begin{vmatrix} b_1 & a_{12} \\ b_2 & a_{22} \end{vmatrix}}{\begin{vmatrix} a_{11} & a_{12} \\ a_{21} & a_{22} \end{vmatrix}} \quad \text{and} \quad x_2 = \frac{\begin{vmatrix} a_{11} & b_1 \\ a_{21} & b_2 \end{vmatrix}}{\begin{vmatrix} a_{11} & a_{12} \\ a_{21} & a_{22} \end{vmatrix}}$$

The solutions to (1) and (2) are seen to be the ratios of determinants. The denominators are the determinants of the coefficients of x_1 and x_2 whilst the numerators are essentially the same determinants with one column, the first for x_1 and the second for x_2, replaced by the column made up of b_1 and b_2. For example,

$$3x_1 - x_2 = 1$$
$$2x_1 + x_2 = 4$$

$$\begin{vmatrix} a_{11} & a_{12} \\ a_{21} & a_{22} \end{vmatrix} = \begin{vmatrix} 3 & -1 \\ 2 & 1 \end{vmatrix} = 3 \times 1 - (-1) \times 2 = 3 + 2 = 5$$

$$\begin{vmatrix} b_1 & a_{12} \\ b_2 & a_{22} \end{vmatrix} = \begin{vmatrix} 1 & -1 \\ 4 & 1 \end{vmatrix} = 1 \times 1 - (-1) \times 4 = 1 + 4 = 5$$

$$\begin{vmatrix} a_{11} & b_1 \\ a_{21} & b_2 \end{vmatrix} = \begin{vmatrix} 3 & 1 \\ 2 & 4 \end{vmatrix} = 3 \times 4 - 1 \times 2 = 12 - 2 = 10$$

Thus the solutions are

$$x_1 = \tfrac{5}{5} = 1, \qquad x_2 = \tfrac{10}{5} = 2$$

It will be clear from this that if the determinant of the coefficients of x_1 and x_2 is zero then there is no unique solution to the equations.

For example,
$$2x_1 + 2x_2 = b_1$$
$$3x_1 + 3x_2 = b_2$$

Here

$$\begin{vmatrix} a_{11} & a_{12} \\ a_{21} & a_{22} \end{vmatrix} = \begin{vmatrix} 2 & 2 \\ 3 & 3 \end{vmatrix} = 2 \times 3 - 2 \times 3 = 6 - 6 = 0$$

and there is no solution. This is because either the two equations are identical and their graphs coincide, or the two equations are contradictory and their graphs are parallel lines. In either case there is no unique point of intersection. Determinants of the kind just discussed, which have two rows and two columns, are known as two by two or (2×2) determinants, and are said to be *second-order* determinants. It is possible to define higher-order determinants such as a (3×3) determinant

$$\begin{vmatrix} a_{11} & a_{12} & a_{13} \\ a_{21} & a_{22} & a_{23} \\ a_{31} & a_{32} & a_{33} \end{vmatrix} = a_{11} \begin{vmatrix} a_{22} & a_{23} \\ a_{32} & a_{33} \end{vmatrix} - a_{12} \begin{vmatrix} a_{21} & a_{23} \\ a_{31} & a_{33} \end{vmatrix} + a_{13} \begin{vmatrix} a_{21} & a_{22} \\ a_{31} & a_{32} \end{vmatrix}$$

which is defined in terms of the three second-order determinants. These are obtained by multiplying a_{11} by the determinant of the coefficients which remains when the first row and first column are eliminated,

$$\begin{vmatrix} a_{11} & a_{12} & a_{13} \\ a_{21} & a_{22} & a_{23} \\ a_{31} & a_{32} & a_{33} \end{vmatrix} \rightarrow \begin{vmatrix} a_{22} & a_{23} \\ a_{32} & a_{33} \end{vmatrix}$$

31

and similarly for the other terms. Further discussion of the properties of determinants is deferred until Chapter 2, and this chapter concludes with the use of third-order determinants in the solution of simultaneous equations.

The general system of three equations in three unknowns can be written

$$a_{11}x_1 + a_{12}x_2 + a_{13}x_3 = b_1$$

$$a_{21}x_1 + a_{22}x_2 + a_{23}x_3 = b_2$$

$$a_{31}x_1 + a_{32}x_2 + a_{33}x_3 = b_3$$

and by analogy with the two-equation system the solution is

$$x_1 = \frac{|\mathbf{A}_1|}{|\mathbf{A}|}, \qquad x_2 = \frac{|\mathbf{A}_2|}{|\mathbf{A}|}, \qquad x_3 = \frac{|\mathbf{A}_3|}{|\mathbf{A}|}$$

where

$$|\mathbf{A}| = \begin{vmatrix} a_{11} & a_{12} & a_{13} \\ a_{21} & a_{22} & a_{23} \\ a_{31} & a_{32} & a_{33} \end{vmatrix}, \qquad |\mathbf{A}_1| = \begin{vmatrix} b_1 & a_{12} & a_{13} \\ b_2 & a_{22} & a_{23} \\ b_3 & a_{32} & a_{33} \end{vmatrix}$$

$$|\mathbf{A}_2| = \begin{vmatrix} a_{11} & b_1 & a_{13} \\ a_{21} & b_2 & a_{23} \\ a_{31} & b_3 & a_{33} \end{vmatrix}, \qquad |\mathbf{A}_3| = \begin{vmatrix} a_{11} & a_{12} & b_1 \\ a_{21} & a_{22} & b_2 \\ a_{31} & a_{32} & b_3 \end{vmatrix}$$

so that $|\mathbf{A}_i|$ is the value of $|\mathbf{A}|$ with column i replaced by the column of b-values. For example,

$$3x_1 + x_2 - 2x_3 = 2$$

$$x_1 + x_2 + x_3 = 1$$

$$2x_1 + 2x_2 + 3x_3 = 3$$

Here

$$|\mathbf{A}| = \begin{vmatrix} 3 & 1 & -2 \\ 1 & 1 & 1 \\ 2 & 2 & 3 \end{vmatrix} = 3 \begin{vmatrix} 1 & 1 \\ 2 & 3 \end{vmatrix} - 1 \begin{vmatrix} 1 & 1 \\ 2 & 3 \end{vmatrix} + (-2) \begin{vmatrix} 1 & 1 \\ 2 & 2 \end{vmatrix}$$

$$= 3(3-2) - 1(3-2) - 2(2-2)$$

$$= 3 \times 1 - 1 \times 1 - 2 \times 0 = 2$$

$$|\mathbf{A_1}| = \begin{vmatrix} 2 & 1 & -2 \\ 1 & 1 & 1 \\ 3 & 2 & 3 \end{vmatrix} = 2\begin{vmatrix} 1 & 1 \\ 2 & 3 \end{vmatrix} - 1\begin{vmatrix} 1 & 1 \\ 3 & 3 \end{vmatrix} + (-2)\begin{vmatrix} 1 & 1 \\ 3 & 2 \end{vmatrix}$$

$$= 2(3-2) - 1(3-3) - 2(2-3) = 4$$

$$|\mathbf{A_2}| = \begin{vmatrix} 3 & 2 & -2 \\ 1 & 1 & 1 \\ 2 & 3 & 3 \end{vmatrix} = 3\begin{vmatrix} 1 & 1 \\ 3 & 3 \end{vmatrix} - 2\begin{vmatrix} 1 & 1 \\ 2 & 3 \end{vmatrix} + (-2)\begin{vmatrix} 1 & 1 \\ 2 & 3 \end{vmatrix}$$

$$= 3(3-3) - 2(3-2) - 2(3-2) = -4$$

$$|\mathbf{A_3}| = \begin{vmatrix} 3 & 1 & 2 \\ 1 & 1 & 1 \\ 2 & 2 & 3 \end{vmatrix} = 3\begin{vmatrix} 1 & 1 \\ 2 & 3 \end{vmatrix} - 1\begin{vmatrix} 1 & 1 \\ 2 & 3 \end{vmatrix} + 2\begin{vmatrix} 1 & 1 \\ 2 & 2 \end{vmatrix}$$

$$= 3(3-2) - 1(3-2) + 2(2-2) = 2$$

Therefore the solution is

$$x_1 = \frac{|\mathbf{A_1}|}{|\mathbf{A}|} = \frac{4}{2} = 2$$

$$x_2 = \frac{|\mathbf{A_2}|}{|\mathbf{A}|} = -\frac{4}{2} = -2$$

$$x_3 = \frac{|\mathbf{A_3}|}{|\mathbf{A}|} = \frac{2}{2} = 1$$

which is easily checked by direct substitution in the original equations.

This method of solving simultaneous equations is known as *Cramer's rule,* and is probably most useful for systems of three equations in three unknowns. For larger systems of equations the arithmetic involved becomes tedious and nowadays computers are generally used to solve simultaneous equations by methods such as matrix inversion, which is explained in the next chapter.

1.10 EXERCISES

Use the determinant method to find the values of the variables which will satisfy the following equations.

1. $2x_1 + 2x_2 = 2$
   $3x_1 - x_2 = 1$

2. $x_1 + 3x_2 = 4$
   $2x_1 + x_2 = 3$

3. $x_1 - x_2 + x_3 = 4$
   $x_1 + x_2 + 3x_3 = 8$
   $x_1 + 2x_2 - x_3 = 0$

4. $2x_1 + 2x_2 + x_3 = 3$
   $2x_1 - 2x_2 + x_3 = -1$
   $x_1 + x_2 - 2x_3 = 4$

5. $x_1 + x_2 - 2x_3 = 0$
   $2x_1 - 2x_2 - 2x_3 = -4$
   $x_1 + 2x_2 + x_3 = 8$

6. $x_1 + x_2 - 2x_3 = 3$
   $2x_1 - 3x_2 + x_3 = 1$
   $x_1 - x_2 - x_3 = 4$

CHAPTER TWO

Elementary Matrix Algebra

2.1 THE MATRIX NOTATION

A *matrix* is a rectangular array of elements arranged in *rows* and *columns*. Examples of matrices are

$$\begin{bmatrix} 1 & 0 \\ 0 & 1 \end{bmatrix} \quad \begin{bmatrix} a & b & c \\ c & c & b \\ 2a & 2c & b \end{bmatrix} \quad \begin{bmatrix} x_{11} & x_{12} & x_{13} \\ x_{21} & x_{22} & x_{23} \end{bmatrix}$$

which are rectangular arrays with 2 rows and 2 columns, 3 rows and 3 columns and 2 rows and 3 columns respectively. The *order* of a matrix is the number of rows and the number of columns, so that in the above examples the orders are (2×2) or (2 by 2), (3×3), and (2×3) respectively. The elements, which may be numbers or constants or variables, are enclosed between square brackets to signify that they must be considered as a whole and not individually. In contrast to the determinants of Chapter 1, a matrix does not have a single numerical value.

A matrix is often denoted by a single letter in bold-face type and a general matrix of order $m \times n$ is written as

$$\mathbf{X} = \begin{bmatrix} x_{11} & x_{12} & \dots & x_{1n} \\ x_{21} & x_{22} & \dots & x_{2n} \\ \vdots & \vdots & & \vdots \\ x_{m1} & x_{m2} & \dots & x_{mn} \end{bmatrix}$$

where the subscripts identify the row and column in which the element is located. For example

x_{12} is the element in row 1, column 2

x_{34} is the element in row 3, column 4

and $\qquad x_{ij}$ is the element in row i, column j.

A *square matrix* has an equal number of rows and columns and in the general case has order $(n \times n)$. If a matrix has only one row it is referred to as a *row vector* and similarly a matrix with only one column is referred to as a *column vector*.

Examples of these are

$$\mathbf{A} = \begin{bmatrix} a_{11} & a_{12} \\ a_{21} & a_{22} \end{bmatrix} \quad \text{is a square matrix of order } (2 \times 2)$$

$$\mathbf{B} = \begin{bmatrix} 3 & 4 & 5 & 1 \end{bmatrix} \quad \text{is a row vector of order } (1 \times 4)$$

$$\mathbf{C} = \begin{bmatrix} c_{11} \\ c_{21} \end{bmatrix} \quad \text{is a column vector of order } (2 \times 1)$$

2.2 ELEMENTARY MATRIX OPERATIONS

Matrix Equality

Two matrices are equal if, and only if, the corresponding elements of each matrix are all equal. For example,

$$\begin{bmatrix} x_1 \\ x_2 \end{bmatrix} = \begin{bmatrix} 4 \\ 2 \end{bmatrix} \qquad \text{if} \qquad x_1 = 4 \quad \text{and} \quad x_2 = 2$$

$$\begin{bmatrix} x_{11} & x_{12} \\ x_{21} & x_{22} \end{bmatrix} = \begin{bmatrix} a_{11} & a_{12} \\ a_{21} & a_{22} \end{bmatrix} \qquad \text{if} \qquad \begin{matrix} x_{11} = a_{11}, & x_{12} = a_{12} \\ x_{21} = a_{21}, & x_{22} = a_{22} \end{matrix}$$

and, in general

$$\begin{bmatrix} x_{11} & x_{12} & \cdots & x_{1n} \\ x_{21} & x_{22} & \cdots & x_{2n} \\ \vdots & \vdots & & \vdots \\ x_{m1} & x_{m2} & \cdots & x_{mn} \end{bmatrix} = \begin{bmatrix} a_{11} & a_{12} & \cdots & a_{1n} \\ a_{21} & a_{22} & \cdots & a_{2n} \\ \vdots & \vdots & & \vdots \\ a_{m1} & a_{m2} & \cdots & a_{mn} \end{bmatrix}$$

that is, $\mathbf{X} = \mathbf{A}$, if $x_{ij} = a_{ij}$ for all i and all j, where x_{ij}, a_{ij} represent the elements in the ith row and jth column of the two matrices respectively.

It follows that two matrices can be equal only if they have the same number of elements arranged in corresponding positions, i.e. they must be of the same order. In the above case, both $\mathbf{X}$ and $\mathbf{A}$ are $(m \times n)$.

SCALAR MULTIPLICATION

A *scalar* is a real number, in contrast to a vector or matrix. If k is a scalar then the product of k and a matrix is obtained by multiplying each element of the matrix by k. For example,

$$\mathbf{X} = \begin{bmatrix} x_{11} & x_{12} \\ x_{21} & x_{22} \\ x_{31} & x_{32} \end{bmatrix} \qquad k\mathbf{X} = \begin{bmatrix} kx_{11} & kx_{12} \\ kx_{21} & kx_{22} \\ kx_{31} & kx_{32} \end{bmatrix}$$

$$\mathbf{A} = \begin{bmatrix} 2 & 4 \\ 0 & -1 \end{bmatrix} \qquad 3\mathbf{A} = \begin{bmatrix} 6 & 12 \\ 0 & -3 \end{bmatrix}$$

ADDITION AND SUBTRACTION

If $\mathbf{A}$ and $\mathbf{B}$ are of the same order then $\mathbf{C} = \mathbf{A} + \mathbf{B}$ is found by adding the corresponding elements in $\mathbf{A}$ and $\mathbf{B}$, and $\mathbf{D} = \mathbf{A} - \mathbf{B}$ is found by subtracting the corresponding elements in $\mathbf{B}$ from $\mathbf{A}$. For example,

Let
$$\mathbf{A} = \begin{bmatrix} 1 & 3 \\ 2 & 4 \\ 0 & -1 \end{bmatrix} \qquad \mathbf{B} = \begin{bmatrix} 2 & x \\ 0 & -1 \\ 2 & 2 \end{bmatrix}$$

then
$$\mathbf{C} = \mathbf{A} + \mathbf{B} = \begin{bmatrix} 1+2 & 3+x \\ 2+0 & 4+(-1) \\ 0+2 & -1+2 \end{bmatrix} = \begin{bmatrix} 3 & 3+x \\ 2 & 3 \\ 2 & 1 \end{bmatrix}$$

and
$$\mathbf{D} = \mathbf{A} - \mathbf{B} = \begin{bmatrix} 1-2 & 3-x \\ 2-0 & 4-(-1) \\ 0-2 & -1-2 \end{bmatrix} = \begin{bmatrix} -1 & 3-x \\ 2 & 5 \\ -2 & -3 \end{bmatrix}$$

37

MATRIX MULTIPLICATION

Two matrices **A** and **B** can be multiplied together to give **AB** only if the number of *columns* in **A** is equal to the number of *rows* in **B**. For example,

if
$$\mathbf{A} = \begin{bmatrix} a_{11} & a_{12} \\ a_{21} & a_{22} \\ a_{31} & a_{32} \end{bmatrix} \quad \text{and} \quad \mathbf{B} = \begin{bmatrix} b_{11} & b_{12} & b_{13} \\ b_{21} & b_{22} & b_{23} \end{bmatrix}$$

then **A** is (3×2), **B** is (2×3) and the product of **A** and **B** can be found since **A** has 2 columns and **B** has 2 rows. The product matrix **C** is

$$\mathbf{C} = \mathbf{AB} = \begin{bmatrix} a_{11} & a_{12} \\ a_{21} & a_{22} \\ a_{31} & a_{32} \end{bmatrix} \begin{bmatrix} b_{11} & b_{12} & b_{13} \\ b_{21} & b_{22} & b_{23} \end{bmatrix}$$

$$= \begin{bmatrix} a_{11}b_{11} + a_{12}b_{21} & a_{11}b_{12} + a_{12}b_{22} & a_{11}b_{13} + a_{12}b_{23} \\ a_{21}b_{11} + a_{22}b_{21} & a_{21}b_{12} + a_{22}b_{22} & a_{21}b_{13} + a_{22}b_{23} \\ a_{31}b_{11} + a_{32}b_{21} & a_{31}b_{12} + a_{32}b_{22} & a_{31}b_{13} + a_{32}b_{23} \end{bmatrix}$$

$$= \begin{bmatrix} c_{11} & c_{12} & c_{13} \\ c_{21} & c_{22} & c_{23} \\ c_{31} & c_{32} & c_{33} \end{bmatrix}$$

where
$$c_{11} = a_{11}b_{11} + a_{12}b_{21}$$
$$c_{12} = a_{11}b_{12} + a_{12}b_{22}$$

and the general term is $c_{ij} = a_{i1}b_{1j} + a_{i2}b_{2j}$.

The elements of **C** are obtained by multiplying row i from **A** by column j from **B** to give c_{ij}.

Notice that **A** is (3×2), **B** is (2×3) and **C** is $(3 \times 2)(2 \times 3) = (3 \times 3)$.

Example 1

$$\mathbf{A} = \begin{bmatrix} 2 & 1 \\ 4 & -1 \end{bmatrix} \quad \mathbf{B} = \begin{bmatrix} 3 \\ 2 \end{bmatrix}$$

A is (2×2), **B** is (2×1) so **C** is $(2 \times 2)(2 \times 1) = (2 \times 1)$

$$\mathbf{C} = \begin{bmatrix} 2 & 1 \\ 4 & -1 \end{bmatrix} \begin{bmatrix} 3 \\ 2 \end{bmatrix} = \begin{bmatrix} 2 \times 3 + 1 \times 2 \\ 4 \times 3 + (-1) \times 2 \end{bmatrix} = \begin{bmatrix} 6 + 2 \\ 12 - 2 \end{bmatrix} = \begin{bmatrix} 8 \\ 10 \end{bmatrix}$$

Example 2

$$A = \begin{bmatrix} 2 & 1 & 4 \\ 3 & 2 & 0 \end{bmatrix}, \qquad B = \begin{bmatrix} 2 & 3 \\ 1 & 4 \end{bmatrix}$$

Here **A** is (2×3) and **B** is (2×2) and so the product **AB** is not defined since the number of columns in **A** does not equal the number of rows in **B**. However, if

$$B = \begin{bmatrix} 2 & 3 \\ 1 & 4 \\ 0 & 1 \end{bmatrix}$$

then **B** is (3×2) and the product is of order $(2 \times 3)(3 \times 2) = (2 \times 2)$

$$
\begin{aligned}
C = AB &= \begin{bmatrix} 2 & 1 & 4 \\ 3 & 2 & 0 \end{bmatrix} \begin{bmatrix} 2 & 3 \\ 1 & 4 \\ 0 & 1 \end{bmatrix} \\
&= \begin{bmatrix} 2 \times 2 + 1 \times 1 + 4 \times 0 & 2 \times 3 + 1 \times 4 + 4 \times 1 \\ 3 \times 2 + 2 \times 1 + 0 \times 0 & 3 \times 3 + 2 \times 4 + 0 \times 1 \end{bmatrix} \\
&= \begin{bmatrix} 5 & 14 \\ 8 & 17 \end{bmatrix}
\end{aligned}
$$

In this case it is also possible to form the product $D = BA$, since **B** is (3×2) and **A** is (2×3) and hence **D** is of order $(3 \times 2)(2 \times 3) = (3 \times 3)$

$$
\begin{aligned}
D = BA &= \begin{bmatrix} 2 & 3 \\ 1 & 4 \\ 0 & 1 \end{bmatrix} \begin{bmatrix} 2 & 1 & 4 \\ 3 & 2 & 0 \end{bmatrix} \\
&= \begin{bmatrix} 2 \times 2 + 3 \times 3 & 2 \times 1 + 3 \times 2 & 2 \times 4 + 3 \times 0 \\ 1 \times 2 + 4 \times 3 & 1 \times 1 + 4 \times 2 & 1 \times 4 + 4 \times 0 \\ 0 \times 2 + 1 \times 3 & 0 \times 1 + 1 \times 2 & 0 \times 4 + 1 \times 0 \end{bmatrix} \\
&= \begin{bmatrix} 13 & 8 & 8 \\ 14 & 9 & 4 \\ 3 & 2 & 0 \end{bmatrix}
\end{aligned}
$$

39

Here $\mathbf{D}$ is (3×3) whilst $\mathbf{C}$ is (2×2). Therefore $\mathbf{D}$ and $\mathbf{C}$ cannot be equal and we have

$$\mathbf{AB} \neq \mathbf{BA}$$

This is true in this particular case and in fact is usually true even when the products $\mathbf{AB}$ and $\mathbf{BA}$ are of the same order.

Example 3

$$\mathbf{A} = \begin{bmatrix} a_{11} & a_{12} \\ a_{21} & a_{22} \end{bmatrix} \qquad \mathbf{X} = \begin{bmatrix} x_1 \\ x_2 \end{bmatrix}$$

$\mathbf{A}$ is (2×2) and $\mathbf{X}$ is (2×1) so the product $\mathbf{AX}$ is $(2 \times 2)(2 \times 1) = (2 \times 1)$ and is

$$\mathbf{AX} = \begin{bmatrix} a_{11} & a_{12} \\ a_{21} & a_{22} \end{bmatrix} \begin{bmatrix} x_1 \\ x_2 \end{bmatrix} = \begin{bmatrix} a_{11}x_1 + a_{12}x_2 \\ a_{21}x_1 + a_{22}x_2 \end{bmatrix}$$

It follows from this that the set of simultaneous equations

$$a_{11}x_1 + a_{12}x_2 = b_1$$

$$a_{21}x_1 + a_{22}x_2 = b_2$$

can be represented by

$$\mathbf{AX} = \mathbf{b}$$

where

$$\mathbf{b} = \begin{bmatrix} b_1 \\ b_2 \end{bmatrix}$$

THE IDENTITY OR UNIT MATRIX

Corresponding to the number 1 in ordinary algebra there is a matrix $\mathbf{I}$, known as the identity or unit matrix, which has the property

$$\mathbf{IA} = \mathbf{A} = \mathbf{AI}$$

(compare with $1 \times a = a = a \times 1$) and $\mathbf{I}$ is the square matrix

$$\mathbf{I} = \begin{bmatrix} 1 & 0 & 0 \\ 0 & 1 & 0 \\ 0 & 0 & 1 \end{bmatrix}$$

for order (3×3) and for order $(n \times n)$,

$$I = \begin{bmatrix} 1 & 0 & \dots & 0 \\ 0 & 1 & \dots & 0 \\ \vdots & \vdots & & \vdots \\ 0 & 0 & \dots & 1 \end{bmatrix}$$

For example, if

$$A = \begin{bmatrix} 1 & 2 \\ 4 & 3 \end{bmatrix}, \qquad I = \begin{bmatrix} 1 & 0 \\ 0 & 1 \end{bmatrix}$$

then

$$IA = \begin{bmatrix} 1 & 0 \\ 0 & 1 \end{bmatrix} \begin{bmatrix} 1 & 2 \\ 4 & 3 \end{bmatrix} = \begin{bmatrix} 1 \times 1 + 0 \times 4 & 1 \times 2 + 0 \times 3 \\ 0 \times 1 + 1 \times 4 & 0 \times 2 + 1 \times 3 \end{bmatrix}$$

$$= \begin{bmatrix} 1 & 2 \\ 4 & 3 \end{bmatrix} = A$$

THE TRANSPOSE OF A MATRIX

If the rows and columns of a matrix are interchanged the new matrix is known as the *transpose* of the original matrix. For example,

if
$$A = \begin{bmatrix} 1 & 2 \\ 3 & 4 \end{bmatrix}$$

the transpose,
$$A' = \begin{bmatrix} 1 & 3 \\ 2 & 4 \end{bmatrix}$$

where the first row of A, 1 2 becomes the first column of A', and the second row of A, 3 4 becomes the second column of A'.

If
$$A = \begin{bmatrix} a_{11} & a_{12} & a_{13} \\ a_{21} & a_{22} & a_{23} \end{bmatrix}$$
then
$$A' = \begin{bmatrix} a_{11} & a_{21} \\ a_{12} & a_{22} \\ a_{13} & a_{23} \end{bmatrix}$$

If
$$A = \begin{bmatrix} a & b & c \end{bmatrix}$$
then
$$A' = \begin{bmatrix} a \\ b \\ c \end{bmatrix}$$

41

2.3 EXERCISES

1. If $\mathbf{A} = \begin{bmatrix} 2 & 4 \\ 1 & 3 \end{bmatrix}$ $\mathbf{B} = \begin{bmatrix} 2 & 0 \\ -1 & 1 \end{bmatrix}$ $\mathbf{C} = \begin{bmatrix} 3 \\ 1 \end{bmatrix}$

   form the following (if they exist)

   (a) $\mathbf{A} + \mathbf{B}$, (b) $\mathbf{A} - 2\mathbf{B}$, (c) $\mathbf{AB}$, (d) $\mathbf{AC}$, (e) $\mathbf{CA}$, (f) $\mathbf{AB}'$, (g) $\mathbf{C}'\mathbf{A}$ and (h) show that $\mathbf{AB} \neq \mathbf{BA}$.

2. Given that

$$\mathbf{A} = \begin{bmatrix} 1 & 0 & 2 \\ 1 & 1 & 3 \\ 0 & 2 & -1 \end{bmatrix} \quad \mathbf{B} = \begin{bmatrix} 2 & 1 \\ 1 & -1 \\ 2 & 2 \end{bmatrix} \quad \mathbf{C} = [1 \quad 0 \quad 2]$$

   form the following (if they exist):

   (a) $\mathbf{AB}$, (b) $\mathbf{A}'\mathbf{B}$, (c) $\mathbf{AC}$, (d) $\mathbf{AC}'$, (e) $\mathbf{CB}$.

2.4 THE INVERSE MATRIX

In Section 2.2 we saw that a set of simultaneous equations could be written in matrix form. For example the equations

$$a_{11}x_1 + a_{12}x_2 + a_{13}x_3 = b_1$$

$$a_{21}x_1 + a_{22}x_2 + a_{23}x_3 = b_2$$

$$a_{31}x_1 + a_{32}x_2 + a_{33}x_3 = b_3$$

can be written as $\mathbf{AX} = \mathbf{b}$

where $\mathbf{A} = \begin{bmatrix} a_{11} & a_{12} & a_{13} \\ a_{21} & a_{22} & a_{23} \\ a_{31} & a_{32} & a_{33} \end{bmatrix}$ $\mathbf{X} = \begin{bmatrix} x_1 \\ x_2 \\ x_3 \end{bmatrix}$ $\mathbf{b} = \begin{bmatrix} b_1 \\ b_2 \\ b_3 \end{bmatrix}$

The solution of these equations is obtained by expressing the unknowns x_1, x_2, x_3 in terms of the known constants a_{11}, a_{12}, ..., a_{33}, b_1, b_2 and b_3. For the matrix equation $\mathbf{AX} = \mathbf{b}$ this is equivalent to transferring the $\mathbf{A}$ to the other side of the equation. To do this we define a matrix $\mathbf{A}^{-1}$, known as the *inverse* of $\mathbf{A}$, which has the property

$$\mathbf{AA}^{-1} = \mathbf{I} = \mathbf{A}^{-1}\mathbf{A}$$

ie the product of a matrix and its inverse is the unit matrix.

Since $$\mathbf{AX} = \mathbf{b}$$

multiplying each side by $\mathbf{A}^{-1}$, we have

$$\mathbf{A}^{-1}\mathbf{AX} = \mathbf{A}^{-1}\mathbf{b}$$

or $$\mathbf{IX} = \mathbf{A}^{-1}\mathbf{b}$$

But $\mathbf{IX} = \mathbf{X}$ and so $$\mathbf{X} = \mathbf{A}^{-1}\mathbf{b}$$

is the solution to the equation.

Before considering how to obtain $\mathbf{A}^{-1}$ from a matrix $\mathbf{A}$ we recall that in Section 1.9 the condition for a solution to exist for a set of simultaneous equations was shown to be that $|\mathbf{A}| \neq 0$, where $|\mathbf{A}|$ is the determinant of the coefficients of the unknowns. In the case of three equations in three unknowns,

$$|\mathbf{A}| = \begin{vmatrix} a_{11} & a_{12} & a_{13} \\ a_{21} & a_{22} & a_{23} \\ a_{31} & a_{32} & a_{33} \end{vmatrix}$$

and this was evaluated as

$$|\mathbf{A}| = a_{11} \begin{vmatrix} a_{22} & a_{23} \\ a_{32} & a_{33} \end{vmatrix} - a_{12} \begin{vmatrix} a_{21} & a_{23} \\ a_{31} & a_{33} \end{vmatrix} + a_{13} \begin{vmatrix} a_{21} & a_{22} \\ a_{31} & a_{32} \end{vmatrix}$$

These (2×2) determinants are the *minors* of the elements of $\mathbf{A}$, so that the minors of a_{11} and a_{12} are respectively

$$\begin{vmatrix} a_{22} & a_{23} \\ a_{32} & a_{33} \end{vmatrix} \quad \text{and} \quad \begin{vmatrix} a_{21} & a_{23} \\ a_{31} & a_{33} \end{vmatrix}$$

and in general the minor of a_{ij} is the determinant of $\mathbf{A}$ with row i and column j eliminated. The *cofactor* (c_{ij}) of the element a_{ij} of the matrix $\mathbf{A}$ is the minor of a_{ij} multiplied by $(-1)^{i+j}$, so that if $i + j$ is even the cofactor and minor are equal, and if $i + j$ is odd, the cofactor is the negative of the minor.

43

For example if

$$\mathbf{A} = \begin{bmatrix} a_{11} & a_{12} & a_{13} \\ a_{21} & a_{22} & a_{23} \\ a_{31} & a_{32} & a_{33} \end{bmatrix}$$

the minor of a_{22} is

$$\begin{vmatrix} a_{11} & a_{12} & a_{13} \\ a_{21} & a_{22} & a_{23} \\ a_{31} & a_{32} & a_{33} \end{vmatrix} = \begin{vmatrix} a_{11} & a_{13} \\ a_{31} & a_{33} \end{vmatrix}$$

and the cofactor of a_{22} is

$$c_{22} = (-1)^{2+2} \begin{vmatrix} a_{11} & a_{13} \\ a_{31} & a_{33} \end{vmatrix} = \begin{vmatrix} a_{11} & a_{13} \\ a_{31} & a_{33} \end{vmatrix}$$

For a_{23} the minor is

$$\begin{vmatrix} a_{11} & a_{12} \\ a_{31} & a_{32} \end{vmatrix}$$

and the cofactor is

$$c_{23} = (-1)^{2+3} \begin{vmatrix} a_{11} & a_{12} \\ a_{31} & a_{32} \end{vmatrix} = - \begin{vmatrix} a_{11} & a_{12} \\ a_{31} & a_{32} \end{vmatrix}$$

It follows that for $\mathbf{A}$ of order 3×3 then

$$|\mathbf{A}| = a_{11}c_{11} + a_{12}c_{12} + a_{13}c_{13}$$

The *cofactor matrix* is the matrix with elements c_{ij}, the cofactors of $\mathbf{A}$.

Example 1

$$\mathbf{A} = \begin{bmatrix} 2 & 1 \\ 0 & 4 \end{bmatrix}$$

The minor of a_{11} is $\quad \begin{vmatrix} 2 & 1 \\ 0 & 4 \end{vmatrix} = 4$

The minor of a_{12} is $\quad \begin{vmatrix} 2 & 1 \\ 0 & 4 \end{vmatrix} = 0$

The minor of a_{21} is $\begin{vmatrix} 2 & 1 \\ 0 & 4 \end{vmatrix} = 1$

The minor of a_{22} is $\begin{vmatrix} 2 & 1 \\ 0 & 4 \end{vmatrix} = 2$

The cofactors are $c_{11} = 4$, $c_{12} = 0$, $c_{21} = -1$, $c_{22} = 2$. The cofactor matrix

is
$$\mathbf{C} = \begin{bmatrix} c_{11} & c_{12} \\ c_{21} & c_{22} \end{bmatrix} = \begin{bmatrix} 4 & 0 \\ -1 & 2 \end{bmatrix}$$

Example 2

$$\mathbf{A} = \begin{bmatrix} 2 & 1 & 1 \\ 0 & 4 & -1 \\ 2 & 2 & 1 \end{bmatrix}$$

The minor of a_{11} is $\begin{vmatrix} 4 & -1 \\ 2 & 1 \end{vmatrix} = 4 + 2 = 6$

The minor of a_{12} is $\begin{vmatrix} 0 & -1 \\ 2 & 1 \end{vmatrix} = 2$

The minor of a_{13} is $\begin{vmatrix} 0 & 4 \\ 2 & 2 \end{vmatrix} = -8$

The minor of a_{21} is $\begin{vmatrix} 1 & 1 \\ 2 & 1 \end{vmatrix} = -1$

The minor of a_{22} is $\begin{vmatrix} 2 & 1 \\ 2 & 1 \end{vmatrix} = 0$

The minor of a_{23} is $\begin{vmatrix} 2 & 1 \\ 2 & 2 \end{vmatrix} = 2$

The minor of a_{31} is $\begin{vmatrix} 1 & 1 \\ 4 & -1 \end{vmatrix} = -5$

The minor of a_{32} is $\begin{vmatrix} 2 & 1 \\ 0 & -1 \end{vmatrix} = -2$

The minor of a_{33} is $\begin{vmatrix} 2 & 1 \\ 0 & 4 \end{vmatrix} = 8$

The cofactor matrix is

$$\mathbf{C} = \begin{bmatrix} 6 & -2 & -8 \\ 1 & 0 & -2 \\ -5 & 2 & 8 \end{bmatrix}$$

One more definition is required before the inverse can be obtained: the *adjoint* of $\mathbf{A}$ is the transpose of the cofactor matrix of $\mathbf{A}$. Thus, if

$$\mathbf{C} = \begin{bmatrix} c_{11} & c_{12} & c_{13} \\ c_{21} & c_{22} & c_{23} \\ c_{31} & c_{32} & c_{33} \end{bmatrix} \qquad \text{adjoint } (\mathbf{A}) = \begin{bmatrix} c_{11} & c_{21} & c_{31} \\ c_{12} & c_{22} & c_{32} \\ c_{13} & c_{23} & c_{33} \end{bmatrix}$$

The inverse of a matrix $\mathbf{A}$ is given by

$$\mathbf{A}^{-1} = \frac{1}{|\mathbf{A}|} \text{ adjoint } (\mathbf{A})$$

This can be shown to be true quite easily when $\mathbf{A}$ is (2×2) and with more difficulty for higher orders.

Let $\quad \mathbf{A} = \begin{bmatrix} a_{11} & a_{12} \\ a_{21} & a_{22} \end{bmatrix} \quad$ then $\quad \mathbf{C} = \begin{bmatrix} a_{22} & -a_{21} \\ -a_{12} & a_{11} \end{bmatrix}$

and hence

$$\text{adjoint } (\mathbf{A}) = \begin{bmatrix} a_{22} & -a_{12} \\ -a_{21} & a_{11} \end{bmatrix}$$

Now

$$|\mathbf{A}| = \begin{vmatrix} a_{11} & a_{12} \\ a_{21} & a_{22} \end{vmatrix} = a_{11}a_{22} - a_{12}a_{21}$$

If $\mathbf{A}^{-1}$ is the inverse then $\mathbf{A}\mathbf{A}^{-1} = \mathbf{I}$.

Now $\quad \mathbf{A}\mathbf{A}^{-1} = \dfrac{\mathbf{A} \text{ adjoint } (\mathbf{A})}{|\mathbf{A}|}$

$$= \frac{1}{|\mathbf{A}|} \begin{bmatrix} a_{11} & a_{12} \\ a_{21} & a_{22} \end{bmatrix} \begin{bmatrix} a_{22} & -a_{12} \\ -a_{21} & a_{11} \end{bmatrix}$$

$$= \frac{1}{|\mathbf{A}|} \begin{bmatrix} a_{11}a_{22} - a_{12}a_{21} & -a_{11}a_{12} + a_{12}a_{11} \\ a_{21}a_{22} - a_{22}a_{21} & -a_{21}a_{12} + a_{22}a_{11} \end{bmatrix}$$

$$= \frac{1}{|\mathbf{A}|} \begin{bmatrix} a_{11}a_{22} - a_{12}a_{21} & 0 \\ 0 & a_{11}a_{22} - a_{12}a_{21} \end{bmatrix}$$

$$= \mathbf{I}$$

Hence $\mathbf{A}^{-1}$ is the inverse of $\mathbf{A}$.

Example 1

$$2x_1 + 4x_2 = 2$$
$$3x_1 + 4x_2 = 1$$

Let

$$\mathbf{A} = \begin{bmatrix} 2 & 4 \\ 3 & 4 \end{bmatrix} \quad \text{then} \quad \mathbf{C} = \begin{bmatrix} 4 & -3 \\ -4 & 2 \end{bmatrix}$$

and $|\mathbf{A}| = -4$

$$\therefore \quad \mathbf{A}^{-1} = \frac{1}{|\mathbf{A}|} \text{ adjoint } (\mathbf{A}) = \frac{1}{-4} \begin{bmatrix} 4 & -4 \\ -3 & 2 \end{bmatrix} = \begin{bmatrix} -1 & 1 \\ \frac{3}{4} & -\frac{1}{2} \end{bmatrix}$$

Check: $\quad \mathbf{A}\mathbf{A}^{-1} = \begin{bmatrix} 2 & 4 \\ 3 & 4 \end{bmatrix} \begin{bmatrix} -1 & 1 \\ \frac{3}{4} & -\frac{1}{2} \end{bmatrix} = \begin{bmatrix} 1 & 0 \\ 0 & 1 \end{bmatrix}$

The solution to the equations is

$$\mathbf{X} = \begin{bmatrix} x_1 \\ x_2 \end{bmatrix} = \mathbf{A}^{-1}\mathbf{b} = \begin{bmatrix} -1 & 1 \\ \frac{3}{4} & -\frac{1}{2} \end{bmatrix} \begin{bmatrix} 2 \\ 1 \end{bmatrix}$$

$$= \begin{bmatrix} -1 \\ 1 \end{bmatrix} \quad \text{and so} \quad x_1 = -1, \quad x_2 = 1$$

Example 2
$$x_1 + x_2 + x_3 = 6$$
$$2x_1 - x_2 + 2x_3 = 6$$
$$x_1 \qquad - x_3 = -2$$

Here

$$\mathbf{A} = \begin{bmatrix} 1 & 1 & 1 \\ 2 & -1 & 2 \\ 1 & 0 & -1 \end{bmatrix} \quad \text{and} \quad \mathbf{C} = \begin{bmatrix} 1 & 4 & 1 \\ 1 & -2 & 1 \\ 3 & 0 & -3 \end{bmatrix}$$

and $|\mathbf{A}| = 6$

$$\mathbf{A}^{-1} = \frac{1}{|\mathbf{A}|} \text{ adjoint } (\mathbf{A}) = \tfrac{1}{6} \begin{bmatrix} 1 & 1 & 3 \\ 4 & -2 & 0 \\ 1 & 1 & -3 \end{bmatrix}$$

Check:

$$\mathbf{A}\mathbf{A}^{-1} = \begin{bmatrix} 1 & 1 & 1 \\ 2 & -1 & 2 \\ 1 & 0 & -1 \end{bmatrix} \tfrac{1}{6} \begin{bmatrix} 1 & 1 & 3 \\ 4 & -2 & 0 \\ 1 & 1 & -3 \end{bmatrix} = \begin{bmatrix} 1 & 0 & 0 \\ 0 & 1 & 0 \\ 0 & 0 & 1 \end{bmatrix}$$

The solution to the equations is

$$\mathbf{X} = \begin{bmatrix} x_1 \\ x_2 \\ x_3 \end{bmatrix} = \mathbf{A}^{-1}\mathbf{b} = \tfrac{1}{6} \begin{bmatrix} 1 & 1 & 3 \\ 4 & -2 & 0 \\ 1 & 1 & -3 \end{bmatrix} \begin{bmatrix} 6 \\ 6 \\ -2 \end{bmatrix} = \begin{bmatrix} 1 \\ 2 \\ 3 \end{bmatrix}$$

and so $x_1 = 1$, $x_2 = 2$, $x_3 = 3$.

2.5 EXERCISES

1. Obtain the inverse of

$$\mathbf{A} = \begin{bmatrix} 2 & 1 \\ 1 & 3 \end{bmatrix}$$

and hence solve the equations

$$2x_1 + x_2 = 4$$
$$x_1 + 3x_2 = 7$$

2. Solve the following equations using the inverse matrix method.

(a) $2x_1 + 5x_2 = 2$
$$3x_1 - 4x_2 = 10$$

(b) $2x_1 + 2x_2 + x_3 = 1$
$$3x_1 + x_2 + x_3 = 2$$
$$x_1 + x_2 + x_3 = 2$$

(c) $x_1 - x_2 + x_3 = 3$
$$2x_1 - 2x_2 + x_3 = 5$$
$$x_1 + x_2 + 2x_3 = 4$$

2.6 LINEAR DEPENDENCE AND RANK

For an inverse to exist it is necessary for the original matrix to be square. This has important implications in the solution of equations because a square matrix of coefficients necessarily means that the number of equations is equal to the number of unknowns. If there are more equations than unknowns then some of these must be *redundant* if a solution exists. This means that some of them provide no additional information that is not already known from the other equations. For example,

$$x + y = 2$$
$$2x + y = 3$$
$$2x + 2y = 4$$

In this case there are three equations in two unknowns, but the third one is simply a multiple of the first and provides no further information. A value for x and for y can be obtained from either the first and second equations or the second and third equations taken together.

49

There is also the possibility that the extra equation is *inconsistent*. For example,

$$x + y = 2$$

$$2x + y = 3$$

$$2x + 2y = 5$$

This set of equations has no solution because the values of x and y obtained by solving the first and second simultaneously do not satisfy the third. In particular, the first and third equations cannot be satisfied simultaneously.

In general, with a set of simultaneous equations, when the number of equations is equal to the number of unknowns a unique solution can be obtained providing that no equation is either redundant or inconsistent with the others. In such a case the equations are said to be *linearly independent* and $|\mathbf{A}| \neq 0$. If they are not then there is no unique solution since those that are redundant can be discarded leaving more unknowns than equations.

For example, $\qquad x_1 + x_2 = 2$

$$2x_1 + 2x_2 = 4$$

These two equations are obviously identical and are linearly dependent. The value of x_1 can only be determined in terms of x_2.

The *rank* of a square matrix is the number of linearly independent rows in the matrix, and its value can be found by carrying out row operations on the matrix. The rules governing the relevant row operations are

1. Any row of the matrix may be divided (or multiplied) by a (non-zero) constant. (This is equivalent to dividing (or multiplying) an equation by a constant).

2. A multiple of any row may be subtracted from (or added to) any other row. (This is equivalent to subtracting (or adding) a multiple of one equation from (to) another, which is the procedure used in solving a set of simultaneous equations by the process of elimination used in Chapter 1.)

The aim of the row operations is to remove redundant equations (ie rows) from the matrix. For example, to determine the rank of the matrix of coefficients of the following equations

$$x_1 + 2x_2 + x_3 = 6$$
$$2x_1 + x_2 + x_3 = 5$$
$$4x_1 + 5x_2 + 3x_3 = 17$$

form the matrix:

$$\begin{bmatrix} 1 & 2 & 1 \\ 2 & 1 & 1 \\ 4 & 5 & 3 \end{bmatrix} \quad \begin{matrix} (1) \\ (2) \\ (3) \end{matrix}$$

From rule 2, subtract twice (1) from row (2)

$$\begin{bmatrix} 1 & 2 & 1 \\ 0 & -3 & -1 \\ 4 & 5 & 3 \end{bmatrix} \quad \begin{matrix} (4) \\ (5) \\ (6) \end{matrix}$$

This gives a 0 on row (5). Now subtract four times row (4) from row (6):

$$\begin{bmatrix} 1 & 2 & 1 \\ 0 & -3 & -1 \\ 0 & -3 & -1 \end{bmatrix} \quad \begin{matrix} (7) \\ (8) \\ (9) \end{matrix}$$

Subtract row (8) from row (9)

$$\begin{bmatrix} 1 & 2 & 1 \\ 0 & -3 & -1 \\ 0 & 0 & 0 \end{bmatrix} \quad \begin{matrix} (10) \\ (11) \\ (12) \end{matrix}$$

This shows that the third equation is redundant and that the rank of the matrix is at the most 2. From rule 1, divide row (11) by -3

$$\begin{bmatrix} 1 & 2 & 1 \\ 0 & 1 & \frac{1}{3} \\ 0 & 0 & 0 \end{bmatrix} \quad \begin{matrix} (13) \\ (14) \\ (15) \end{matrix}$$

Subtracting twice row (14) from row (13):

$$\begin{bmatrix} 1 & 0 & \frac{1}{3} \\ 0 & 1 & \frac{1}{3} \\ 0 & 0 & 0 \end{bmatrix} \quad \begin{matrix} (16) \\ (17) \\ (18) \end{matrix}$$

51

It is not possible to reduce the matrix any further since the second column has a 1 on row (17) and zeros elsewhere. Thus there are two independent rows and the rank of the matrix is 2.

Example
$$x_1 + 2x_2 + 2x_3 = 7$$
$$x_1 - x_2 + x_3 = 4$$
$$x_1 + x_2 + 3x_3 = 6$$

The matrix of the coefficients is

$$\begin{bmatrix} 1 & 2 & 2 \\ 1 & -1 & 1 \\ 1 & 1 & 3 \end{bmatrix} \quad \begin{matrix} (1) \\ (2) \\ (3) \end{matrix}$$

Subtract row (1) from row (2) and row (3)

$$\begin{bmatrix} 1 & 2 & 2 \\ 0 & -3 & -1 \\ 0 & -1 & 1 \end{bmatrix} \quad \begin{matrix} (4) \\ (5) \\ (6) \end{matrix}$$

Add twice row (5) to row (4) and row (6) to row (5)

$$\begin{bmatrix} 1 & -4 & 0 \\ 0 & -4 & 0 \\ 0 & -1 & 1 \end{bmatrix} \quad \begin{matrix} (7) \\ (8) \\ (9) \end{matrix}$$

Subtract row (8) from row (7), divide row 8 by -4.

$$\begin{bmatrix} 1 & 0 & 0 \\ 0 & 1 & 0 \\ 0 & -1 & 1 \end{bmatrix} \quad \begin{matrix} (10) \\ (11) \\ (12) \end{matrix}$$

Add row (11) to row (12)

$$\begin{bmatrix} 1 & 0 & 0 \\ 0 & 1 & 0 \\ 0 & 0 & 1 \end{bmatrix} \quad \begin{matrix} (13) \\ (14) \\ (15) \end{matrix}$$

This matrix cannot be reduced further, so all the equations are independent and the rank is 3.

2.7 EXERCISES

1. Determine the rank of the following matrices.

(a) $\begin{bmatrix} 1 & 1 & 3 \\ 2 & 1 & 1 \\ 1 & -2 & -1 \end{bmatrix}$ (b) $\begin{bmatrix} 2 & 2 & -4 \\ 2 & -1 & -1 \\ 1 & 1 & -2 \end{bmatrix}$

(c) $\begin{bmatrix} 1 & 0 & 0 \\ 1 & 1 & 0 \\ 0 & 2 & 1 \end{bmatrix}$ (d) $\begin{bmatrix} 2 & 1 & 1 \\ 1 & 0 & 2 \\ 2 & 2 & -1 \end{bmatrix}$

2. Test whether the following equations have a unique solution.

$$2x_1 + 3x_2 - x_3 + 2x_4 = 2$$

$$x_1 - x_2 + 2x_3 + x_4 = 3$$

$$x_1 + x_2 + x_3 - 2x_4 = 3$$

$$2x_1 + 2x_2 - 2x_3 + x_4 = 1$$

2.8 INPUT–OUTPUT ANALYSIS

INTRODUCTION

In any economy there are a number of industries supplying consumer demand. Many of these industries also supply intermediate products which are further processed or utilised by other industries before reaching the final consumer. For example, the glass industry supplies finished products such as mirrors, but also supplies window glass to the building industry to be used in the construction of houses, and toughened glass to the motor industry to be used as windscreens for cars.

The industrial sector of any economy, therefore, consists of a number of interlinked sections from which final consumer demand is met. Changes in demand for any one of the final outputs affect the outputs required from many sections of the economy. In order to see the effects that such changes might produce it is necessary to build a model of the system. This was first attempted by Leontief and the theory has been developed under the name *input–output analysis*.

In building the model Leontief made an important assumption: that the output of one industry which is required to satisfy the demand from another industry is directly proportional to the final output of the latter. This can be written as

$$x_{ij} = a_{ij} X_j$$

where X_j is the total output of industry j,

a_{ij} is a constant of proportionality, and is the proportion of output of industry i that will be required by industry j in order to produce one unit of output. This will depend upon the technology of industry j,

x_{ij} is the output of industry i that will be required by industry j.

For example, let us assume then that the average number of units of glass which is required in the construction of a house is equal to 2. Then this will be the constant of proportionality and the relationship can be expressed as

$$x_{ij} = a_{ij} X_j = 2X_j$$

where X_j is the output of the building industry in terms of the number of house units constructed, and x_{ij} is the output of the glass industry which is required by the building industry for house construction. These figures are, in fact, flows and represent the output per unit of time, eg output per year.

If the technology of the industry changes and houses are constructed with much larger expanses of glass or the average size of a house unit increases, then a new constant of proportionality must be used: for example,

$$x_{ij} = 3X_j$$

The constant of proportionality used in this example is an average for all types of houses. It may be possible, and it would certainly be useful, to classify the different class of dwellings by size and by type, eg bungalows, semi-detached, detached, etc., and to use a separate constant of proportionality for each sub-category.

An industry may also use some of its own output for further processing or as raw material input and as this demand must be met, it must be included as

$$x_{jj} = a_{jj}X_j$$

In this case the constant of proportionality must be less than one, otherwise the industry would require a greater amount of input of its own output than it was capable of producing. (It is possible to consider the system in terms of net outputs only and to ignore these inputs altogether, but this is not done here.)

The total output of any industry depends upon the demands made upon it by all other industries and also upon the demand for its products made directly by the consumer. This latter, the final demand, is a function of consumers' preferences, relative prices and consumers' income, and also varies with time.

For any economy, it is possible to collect data for the flow of goods and services between industries and to specify the final demand for the outputs at any given point in time. In the simple case of a two-industry economy the information relating to the flow of goods which is required to produce a particular level of output might be as presented in Table 2.1.

TABLE 2.1

	Input to Industry 1	Input to Industry 2	Level of output
Industry 1	$x_{11} = 200$	$x_{12} = 400$	$X_1 = 1,600$
Industry 2	$x_{21} = 600$	$x_{22} = 100$	$X_2 = 2,700$

From this data it is possible to establish the technological coefficients a_{ij} using the formula quoted earlier:

$$x_{ij} = a_{ij}X_j$$

therefore

$$a_{ij} = \frac{x_{ij}}{X_j}$$

55

TABLE 2.2

| | Technological coefficient for | |
	Industry 1	Industry 2
Industry 1	$\dfrac{200}{1,600} = 0.125$	$\dfrac{400}{2,700} = 0.148$
Industry 2	$\dfrac{600}{1,600} = 0.375$	$\dfrac{100}{2,700} = 0.037$

From the information given in Table 2.1 it is also possible to determine the final demand which can be supplied. This is given by the level of output less the amounts which are required to meet the demands made by the two industries in producing these outputs. Thus, level of final demand which can be met by Industry 1 is equal to

$$1,600 - (200 + 400) = 1,000 \text{ units}$$

and for Industry 2 it is equal to

$$2,700 - (600 + 100) = 2,000 \text{ units}$$

From the information given in Table 2.2 it is possible to determine the final demand which can be met at any other level of output within the range over which the constants of proportionality can be assumed to hold. For example, the information in Table 2.2 tells us that Industry 1 required 0.125 units of its own output to be used as input for each further unit of output. Also Industry 2 requires 0.148 units of the output of Industry 1 for each unit of its own output.

It follows that, in general, if X_1 and X_2 represent the total output of Industries 1 and 2 respectively and C_1 and C_2 are the final demands for the outputs of the industries respectively we can write

$$0.125\, X_1 + 0.148\, X_2 + C_1 \leqslant X_1$$

and $\qquad 0.375\, X_1 + 0.037\, X_2 + C_2 \leqslant X_2$

The first equation states that the total demand for the product of Industry 1 must be less than or equal to the total output of Industry 1. The second equation states the same thing for Industry 2.

If we assume that the total output of each industry is just sufficient to meet all the demands made upon it the inequalities can be replaced by equations as follows:

$$0.125\ X_1 + 0.148\ X_2 + C_1 = X_1$$

$$0.375\ X_1 + 0.037\ X_2 + C_2 = X_2$$

These equations can then be rearranged

$$X_1 - 0.125\ X_1 - 0.148\ X_2 = C_1$$

$$X_2 - 0.375\ X_1 - 0.037\ X_2 = C_2$$

that is

$$(1 - 0.125)\ X_1 - 0.148\ X_2 = C_1$$

and

$$-0.375\ X_1 + (1 - 0.037)\ X_2 = C_2$$

A pair of simultaneous equations such as these can be conveniently written in matrix notation, as was shown in the earlier part of this chapter.

$$\begin{bmatrix} (1-0.125) & -0.148 \\ -0.375 & (1-0.037) \end{bmatrix} \begin{bmatrix} X_1 \\ X_2 \end{bmatrix} = \begin{bmatrix} C_1 \\ C_2 \end{bmatrix}$$

The condition that this set of equations should have a unique solution in terms of X_1 and X_2 is that the determinant of the matrix of coefficients is not equal to zero.

In this particular type of problem it is necessary to apply the further condition that the values for X_1 and X_2, the output of the industries, must not be negative. This condition is satisfied if the value of the determinant is positive. That is,

$$\begin{vmatrix} (1-0.125) & -0.148 \\ -0.375 & (1-0.037) \end{vmatrix} > 0$$

If this condition holds it is possible to use the equations to determine either

(a) The final demands which can be met when the output of each industry is fully utilised, ie C_1 and C_2 given X_1 and X_2

(b) the total outputs which are required to meet a given level of final demand, ie X_1 and X_2 given C_1 and C_2.

For example, if the total output of the two industries is 1,600 and 2,700 units respectively, the maximum final demand can be obtained from the following relationships:

$$\begin{bmatrix} 0.125 & 0.148 \\ 0.375 & 0.037 \end{bmatrix} \begin{bmatrix} 1,600 \\ 2,700 \end{bmatrix} + \begin{bmatrix} C_1 \\ C_2 \end{bmatrix} = \begin{bmatrix} 1,600 \\ 2,700 \end{bmatrix}$$

or

$$\begin{bmatrix} C_1 \\ C_2 \end{bmatrix} = \begin{bmatrix} 1-0.125 & -0.148 \\ -0.375 & 1-0.037 \end{bmatrix} \begin{bmatrix} 1,600 \\ 2,700 \end{bmatrix}$$

$$= \begin{bmatrix} 0.875 \times 1,600 - 0.148 \times 2,700 \\ -0.375 \times 1,600 + 0.963 \times 2,700 \end{bmatrix}$$

$$= \begin{bmatrix} 1,400 - 400 \\ -600 + 2,600 \end{bmatrix} = \begin{bmatrix} 1,000 \\ 2,000 \end{bmatrix}$$

therefore $C_1 = 1,000$ and $C_2 = 2,000$ which agrees with the information given in Table 2.1.

Let us now assume that the level of final demand for the output of the two industries is reversed, ie $C_1 = 2,000$ units and $C_2 = 1,000$ units.

Then we can calculate the levels of output which would be required to meet these final demands from the following relationship:

$$\begin{bmatrix} 0.875 & -0.148 \\ -0.375 & 0.963 \end{bmatrix} \begin{bmatrix} X_1 \\ X_2 \end{bmatrix} = \begin{bmatrix} 2,000 \\ 1,000 \end{bmatrix}$$

The inverse of the coefficient matrix is

$$\frac{1}{0.7871} \begin{bmatrix} 0.963 & 0.148 \\ 0.375 & 0.875 \end{bmatrix}$$

and hence

$$\begin{bmatrix} X_1 \\ X_2 \end{bmatrix} = \frac{1}{0.7871} \begin{bmatrix} 0.963 & 0.148 \\ 0.375 & 0.875 \end{bmatrix} \begin{bmatrix} 2,000 \\ 1,000 \end{bmatrix} = \begin{bmatrix} 2,635 \\ 2,064 \end{bmatrix}$$

ie the required output of Industry 1 is 2,635 units and of Industry 2 is 2,064 units (approximately).

2.9 THE INPUT–OUTPUT MATRIX

In general the technology matrix for a two-sector economy can be written as

$$\begin{bmatrix} a_{11} & a_{12} \\ a_{21} & a_{22} \end{bmatrix}$$

or simply as **A**. This matrix is known as the *input–output matrix*. To satisfy final demand exactly from current output the following equality must be true:

$$\begin{bmatrix} a_{11} & a_{12} \\ a_{21} & a_{22} \end{bmatrix} \begin{bmatrix} X_1 \\ X_2 \end{bmatrix} + \begin{bmatrix} C_1 \\ C_2 \end{bmatrix} = \begin{bmatrix} X_1 \\ X_2 \end{bmatrix}$$

or
$$\mathbf{AX} + \mathbf{C} = \mathbf{X}$$

These equations can be rearranged in the form

$$\begin{bmatrix} (1-a_{11}) & -a_{12} \\ -a_{21} & (1-a_{22}) \end{bmatrix} \begin{bmatrix} X_1 \\ X_2 \end{bmatrix} = \begin{bmatrix} C_1 \\ C_2 \end{bmatrix}$$

or
$$(\mathbf{I} - \mathbf{A})\mathbf{X} = \mathbf{C}$$

where **I** is the unit matrix and $(\mathbf{I} - \mathbf{A})$ is known as the *Leontief matrix*. For an economy which is divided into n sectors

$$\mathbf{A} = \begin{bmatrix} a_{11} & a_{12} \cdots a_{1n} \\ a_{21} & a_{22} \cdots a_{2n} \\ \vdots & \vdots \quad \vdots \\ a_{n1} & a_{n2} \cdots a_{nn} \end{bmatrix}$$

$$(\mathbf{I} - \mathbf{A}) = \begin{bmatrix} (1-a_{11}) & -a_{12} & \cdots & -a_{1n} \\ -a_{21} & (1-a_{22}) & \cdots & -a_{2n} \\ \vdots & & & \\ -a_{n1} & -a_{n2} & \cdots & (1-a_{nn}) \end{bmatrix}$$

The output of each industry which is required to satisfy a given final demand from current production can be found by inverting this matrix.

The changes in output which would be required to meet different sets of final demand can be found by multiplying the inverse of the input–output matrix by the vectors which represent the different sets of final demand.

For example, Table 2.3 is the technological matrix for a four-sector economy.

TABLE 2.3

Absorbing sector Producing sector	1	2	3	4	Final demand
1	0.269	0.219	0.246	0.062	3,662
2	0.010	0.239	0.200	0.076	1,259
3	0.008	0.006	0.051	0.230	1,248
4	0.015	0.154	0.074	0.720	4,789

From this matrix,

$$(\mathbf{I} - \mathbf{A}) = \begin{bmatrix} (1-0.269) & -0.219 & -0.246 & -0.062 \\ -0.010 & (1-0.239) & -0.200 & -0.076 \\ -0.008 & -0.006 & (1-0.051) & -0.230 \\ -0.015 & -0.154 & -0.074 & (1-0.720) \end{bmatrix}$$

and

$$(\mathbf{I} - \mathbf{A})^{-1} = \begin{bmatrix} 1.401 & 0.597 & 0.562 & 0.934 \\ 0.039 & 1.474 & 0.377 & 0.718 \\ 0.038 & 0.233 & 1.195 & 1.053 \\ 0.107 & 0.904 & 0.553 & 4.295 \end{bmatrix}$$

The output of each industry which is required to satisfy the final demand is found by multiplying the inverse matrix by the given final demand vector.

$$\mathbf{X} = (\mathbf{I} - \mathbf{A})^{-1}\mathbf{C} = \begin{bmatrix} 1.401 & 0.597 & 0.562 & 0.934 \\ 0.039 & 1.474 & 0.377 & 0.718 \\ 0.038 & 0.233 & 1.195 & 1.053 \\ 0.107 & 0.904 & 0.553 & 4.295 \end{bmatrix} \begin{bmatrix} 3662 \\ 1259 \\ 1248 \\ 4789 \end{bmatrix}$$

$$= \begin{bmatrix} 11,056 \\ 5,908 \\ 6,967 \\ 22,789 \end{bmatrix}$$

$$\therefore \quad x_1 = 11,056, \quad x_2 = 5,908, \quad x_3 = 6,967, \quad x_4 = 22,789$$

The conditions that must be satisfied in order that at least one set of final demands can be met from any given set of output levels are

(a)

$$\begin{vmatrix} (1-a_{11}) & -a_{12} & \cdots & -a_{1n} \\ -a_{21} & (1-a_{22}) & \cdots & -a_{2n} \\ \vdots & \vdots & & \\ -a_{n1} & -a_{n2} & \cdots & (1-a_{nn}) \end{vmatrix} > 0$$

and

(b) all the elements on the main diagonal of the matrix, namely

$(1-a_{11}), (1-a_{22}), \ldots, (1-a_{nn})$, are positive; that is,

$$(1-a_{11}) > 0, \ldots, (1-a_{ii}) > 0, \ldots, (1-a_{nn}) > 0$$

These are known as the *Hawkins–Simon* conditions.

It is possible to approach the problem slightly differently and to consider the total available capacity of each industry. Let x_i represent the maximum output which can be obtained from industry i. Then we can calculate the value of the vector **C**, (the maximum final demands for each industry), which could be met if every industry produced at its maximum level. This would be obtained from the relationship

$$\mathbf{C} = (\mathbf{I} - \mathbf{A})\mathbf{X}$$

However, we often find that the result is a **C**-vector containing one or more negative elements. These would correspond to a negative final demand which is impossible and means that the economy would not operate with all its industries at their maximum output level.

The problem now becomes one of deciding which industries should operate at less than their maximum output, ie to decide where the 'slack' capacity should be. The latter can only be decided in terms of some specified objective. For example, let C_1 be the final demand for the output of industry i which is to be satisfied at a price p_1 and in general let C_n be the final demand for the output of industry n which is to be satisfied at a price p_n. One objective may then be stated as:

Maximise

$$Z = C_1 p_1 + C_2 p_2 + \cdots + C_n p_n = \sum_{i=1}^{n} C_i p_i$$

subject to the following constraints on the system

$$a_{11} X_1 + a_{12} X_2 + \cdots + a_{1n} X_n + C_1 \leqslant X_1$$
$$a_{21} X_1 + a_{22} X_2 + \cdots + a_{2n} X_n + C_2 \leqslant X_2$$
$$\vdots$$
$$a_{n1} X_1 + a_{n2} X_2 + \cdots + a_{nn} X_n + C_n \leqslant X_n$$

or $\qquad\qquad \mathbf{AX + C \leqslant X}$

This is, in fact, a linear-programming type of problem and the method of arriving at an optimum solution in such situations is discussed in Chapter 8.

2.10 EXERCISES

1. Determine the maximum final demand which can be met in the following situation shown in Table 2.4.

TABLE 2.4

	Input to		Level of output
	Industry 1	Industry 2	
Industry 1	200	300	1,500
Industry 2	500	100	2,500

2. What final demand can be met when the level of output of Industry 1 is increased to 2,000 units in a situation which is in all other respects identical to that given in Question 1?

3. Determine the level of output which is necessary to meet final demands of 1,000 and 2,000 respectively when the technological coefficients are given by the following matrices:

(a) $\begin{bmatrix} 0.2 & 0.4 \\ 0.3 & 0.2 \end{bmatrix}$ (b) $\begin{bmatrix} 0 & 0.6 \\ 0.4 & 0 \end{bmatrix}$

CHAPTER THREE

Non-Linear Equations

3.1 THE QUADRATIC

The relationship between two variables can be expressed as $y = f(x)$ and in Chapter 1 we considered the case where the function was linear in x and the equation could be written as

$$y = a + bx$$

In the production situation such an equation implies that the variable cost is constant for all levels of output within the range for which the equation is valid.

In many practical cases it is not possible to make such an assumption and it is obvious that a linear relationship will not be an adequate representation of the situation. However it may be possible to make use of the *quadratic function*, for which the general form is as follows:

$$y = ax^2 + bx + c$$

where a, b and c are constants.

This is an equation of degree two because it contains a term in x^2 but no terms in x^3 or higher powers. An extra term has been added to the general linear equation $y = a + bx$ and the effect of this is to give the graph of the quadratic a curved form. The direction of the curvature and the amount it deviates from the straight line are determined by the sign and size of the constant a.

This can be seen by determining whether a quadratic equation fits the following cost information.

Outputs (units) x	10	11	50	51
Total cost (£) y	135	158	2615	2718

The general quadratic equation is

$$y = ax^2 + bx + c$$

This contains three constants a, b and c, and to determine the values of these constants requires three sets of independent data. In this case there are four sets of data available, but we shall only use the first three sets and then determine whether the equation fits the fourth set.

Substituting pairs of values of y and x in the general quadratic gives

$$135 = a(10)^2 + b(10) + c$$
$$158 = a(11)^2 + b(11) + c$$
$$2615 = a(50)^2 + b(50) + c$$

Or,
$$135 = 100a + 10b + c \tag{1}$$
$$158 = 121a + 11b + c \tag{2}$$
$$2615 = 2500a + 50b + c \tag{3}$$

These three equations can be solved for the three unknowns a, b and c using Cramer's rule (see Chapter 1) to give $a = 1$, $b = 2$, $c = 15$.

The cost function can then be represented by

$$y = 15 + 2x + x^2$$

This can be checked by reference to the original cost data. When $x = 51$ we know that the cost is £2,718. By substituting $x = 51$ we have

$$y = 15 + 2(51) + 51^2$$
$$= 15 + 102 + 2,601$$
$$= 2,718$$

which agrees with the fourth piece of information which is provided.

If we assume that costs vary continuously and uniformly with output, it is possible to represent the cost function graphically as shown in Fig 3.1.

x	0	10	20	30	40	50
$2x$	0	20	40	60	80	100

x^2	0	100	400	900	1,600	2,500
y	15	135	455	975	1,695	2,615

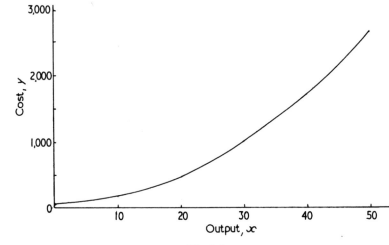

Fig 3.1

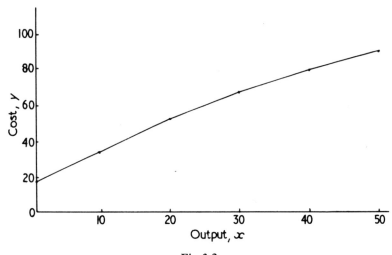

Fig 3.2

Other processes may be represented by different quadratic functions, in particular one in which the coefficient of x^2 is negative. For example,

$$y = 15 + 2x - \frac{x^2}{100}$$

In this case the graph of the function is as shown in Fig 3.2.

x	0	10	20	30	40	50
$2x$	0	20	40	60	80	100
$-x^2/100$	0	-1	-4	-9	-16	-25
y	15	34	51	66	79	90

3.2 EXERCISES

Determine the equations of the quadratic cost functions which fit the following data. Sketch the graphs for the range $x = 0$ to $x = 6$.

1	Output	0	2	6
	Cost	4	14	58
2	Output	4	6	8
	Cost	26	40	58
3	Output	4	6	8
	Cost	31	37	45
4	Output	0	5	10
	Cost	20	24.5	30
5	Output	0	2	4
	Cost	20	22	28

3.3 AVERAGE COST

The average cost of production can be obtained by dividing the total cost by the quantity of output:

$$\text{Average cost} = \frac{\text{total cost}}{\text{quantity of output}}$$

67

Notice that this is an *identity*, ie, it is always true, since it is the definition of average cost.

If the cost function is linear we have

$$\text{Total cost } (TC) = a + bx$$

where x is the quantity of output

$$\therefore \quad \text{Average cost } (AC) = \frac{TC}{x} = \frac{a+bx}{x}$$

$$\therefore \quad AC = \frac{a}{x} + b$$

This is an equation which can be represented graphically if we know the value of the constants. For example,

If
$$TC = 100 + 3x$$

then
$$AC = \frac{100}{x} + 3$$

Values of average cost at various levels of output are obtained in the usual way (with only positive values of x being considered).

$$x = \quad 1 \quad 10 \quad 20 \quad 50 \quad 100$$

$$\frac{100}{x} = 100 \quad 10 \quad 5 \quad 2 \quad 1$$

$$\therefore \quad AC = 103 \quad 13 \quad 8 \quad 5 \quad 4$$

and the graph is as shown in Fig 3.3.

This curve is a rather special one. It is known as a *rectangular hyperbola*. We can note that in this case there is no value of average cost corresponding to the value $x = 0$. This seems obvious because there can be no average cost if there is no output. Economically, the value of x is expected to be greater than 0. Mathematically, if we consider smaller and smaller values of x including fractional values, eg 1/2, 1/4, 1/10, 1/100, we can see that the value of the function becomes larger and larger but, in fact, it never actually touches the vertical axis in a finite point. In such a case we say that the function

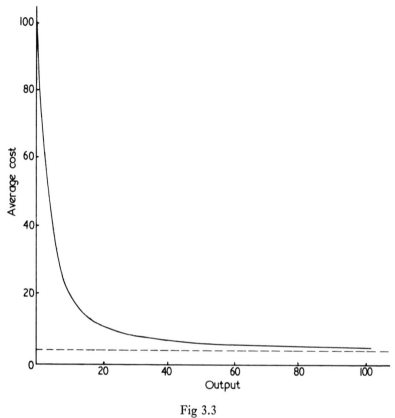

Fig 3.3

tends to a very large value, generally known as *infinity*, as x tends to the value zero. This can be written in mathematical shorthand using the following notation:

$$\text{limit } f(x) \to \infty \qquad \text{as} \qquad x \to 0$$

where $f(x) = 100/x + 3$ and the arrow $\to$ indicates that the function tends to the value ∞ as x tends to the value 0 but never actually attains it.

As output increases the average cost of production falls continuously. This is because the variable cost is constant, but the fixed cost is now

69

spread over more and more units of output. The average cost must, therefore, continue to fall as output increases, but it cannot ever reach a value as low as 3. This is because $AC = 100/x + 3$ and although $100/x$ becomes successively smaller, it never becomes negative, and therefore, even at exceedingly large values of output the average cost is close to but greater than the value 3. In this case we say that as output x tends to the value infinity the average cost tends to the value 3.

$$\text{limit } f(x) = 3 \qquad \text{as} \qquad x \to \infty$$

or more concisely

$$\lim_{x \to \infty} f(x) = 3$$

The curve and the straight line $y = 3$ therefore become closer and closer together as x increases, but they never actually meet. In such

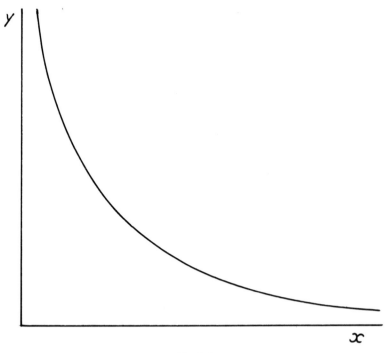

Fig 3.4

cases the curve is said to approach the line *asymptotically*, and the line $y = 3$ is known as an *asymptote*. The y axis (ie the line $x = 0$) is also an asymptote since the curve never actually meets this axis. The equation graphed in Fig 3.4

$$y = \frac{100}{x}$$

is that of a rectangular hyperbola. In this case the lines $x = 0$ and $y = 0$ are the asymptotes and the equation can be written as

$$xy = 100$$

or, in the general case,

$$xy = c$$

where c is a constant.

The equation $xy = c$ describes a whole family of curves which are all rectangular hyperbolae and in which each individual curve is determined by the value of c. The graphs of some such curves are shown in Fig 3.5, where c takes on the successive values 10, 20, 50, 100.

Between certain limits this type of curve may be a good representation of a demand function because if the quantity demanded of a good is inversely proportional to its price then

$$q_d \propto \frac{1}{p}$$

or $q_d p = c$, where c is a constant. We know that total revenue is equal to the product of the quantity demanded and the price of the good. Therefore, a company which has a demand function which can be represented by a rectangular hyperbola $q_d p = c$ has a total revenue which is independent of the quantity sold, ie total revenue is constant.

3.4 BREAKEVEN POINT

This was defined in Chapter 1 as that output at which the total cost and the total revenue are equal and, in general, a unique value can be obtained when the cost and revenue functions are both linear. But this is not necessarily true when one or both of the functions are non-linear.

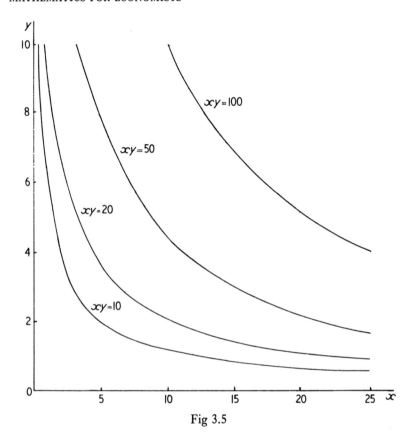

Fig 3.5

Let us first consider the case in which the revenue function is linear, ie the price is independent of the quantity sold, and the cost function is quadratic. For example,

$$\text{Cost function } y_c = 7 + 2x + x^2$$

$$\text{Revenue function } y_R = 10x$$

Then at the breakeven point

$$y_c = y_R$$

$$\therefore \quad 7 + 2x + x^2 = 10x$$

72

or
$$x^2 - 8x + 7 = 0$$

This quadratic equation can be split into factors as follows:

$$(x - 7)(x - 1) = 0$$

and, therefore, either $x = 1$ or $x = 7$ satisfies the equation. This means that there are two points at which total revenue equals total cost and this can be seen to be true by sketching both graphs on the same set of axes as shown in Fig 3.6.

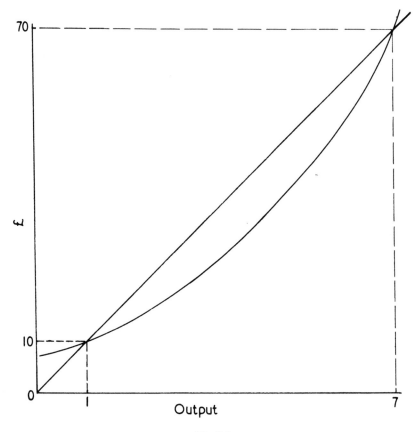

Fig 3.6

In this case net revenue does not increase continuously with output but has the following pattern:

$x < 1$ Net revenue is negative

$x = 1$ Net revenue is zero

$1 < x < 7$ Net revenue increases at first and then decreases again

$x = 7$ Net revenue is zero

$x > 7$ Net revenue is negative

It therefore has a maximum value for an output somewhere between $x = 1$ and $x = 7$. The output at which this maximum value occurs can be obtained using the differential calculus as we shall see later in Chapter 5.

For the present it is sufficient to note that the linear equation and the quadratic equation intersect in two points and the maximum net revenue occurs at an output somewhere between these two points.

But this is not always so. Consider the case where the following equations apply:

$$\text{Cost function } y_c = 16 + 2x + x^2$$

$$\text{Revenue function } y_R = 10x$$

Then $y_c = y_R$

when $16 + 2x + x^2 = 10x$

that is, when $x^2 - 8x + 16 = 0$

This can be factorised as follows:

$$(x - 4)(x - 4) = 0$$

and the solution is simply $x = 4$ which is shown in Fig 3.7.

In Fig 3.6 there are two different points of intersection, whereas in Fig 3.7 these two points coincide, so that the straight line touches rather than intersects the curve. In such a case it is known as the *tangent* to the curve.

Mathematically $x = 4$, $x = 4$ are the two solutions.

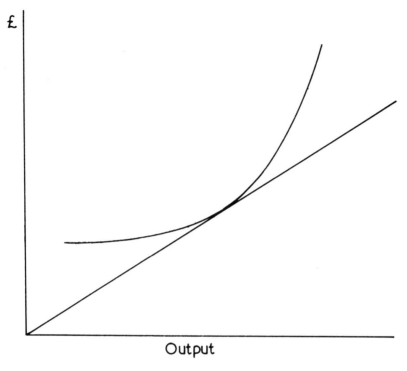

Fig 3.7

Economically, it is apparent that there is only one point at which the net revenue is non-negative and at no point is it positive. It is therefore likely that the company will consider either of the two following possibilities.

(a) A price increase if this is possible. The resulting revenue function would then cut the cost function in two places and if the necessary output could be sold at the new price a positive net revenue could be obtained.

(b) A cost reduction in the process. This would again result in the cost and revenue functions intersecting in two points between which there would be an area of positive net revenue.

75

One further case can occur when the two functions do not intersect at all. This is illustrated graphically in Fig 3.8.

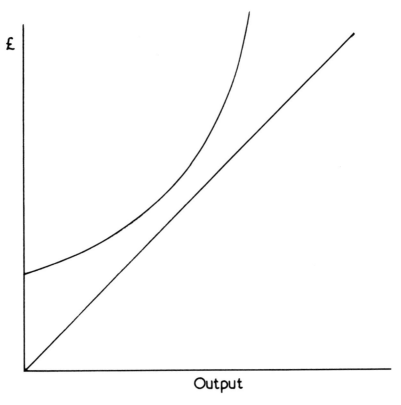

Output

Fig 3.8

In this case there is no output for which there is a positive net revenue. This situation would arise if the fixed costs of the plant were to increase without a corresponding increase in the price of the product. For example,

$$\text{Cost function } y_c = 20 + 2x + x^2$$
$$\text{Revenue function } y_R = 10x$$

Then $$y_c = y_R$$

when $$20 + 2x + x^2 = 10x$$

that is, when $$x^2 - 8x + 20 = 0$$

This equation has no factors, or more precisely this equation has no *real* factors. By a real factor we mean a factor which is an 'ordinary' number such as -6 or 4.68. In the next section we will introduce numbers which are known as *imaginary* numbers, and these will enable us to obtain factors for any quadratic equation.

3.5 ROOTS OF A QUADRATIC EQUATION

How do we know that the quadratic equation

$$x^2 - 8x + 20 = 0$$

has no real roots? Generally speaking we re-arrange the coefficients in the equation to obtain the factors but in many cases this can be difficult. This is not, however, necessary because there is a simple formula which can be used to determine the roots of any quadratic equation. The formula can be derived as follows

Let $ax^2 + bx + c = 0$ be the quadratic equation. Dividing throughout by the coefficient of x^2 and re-arranging the equation

$$x^2 + \frac{b}{a}x + \frac{c}{a} = 0$$

the terms containing x can be made into a perfect square by adding the term $(b/2a)^2$ and if we also subtract this term from the left-hand side, the equation is unchanged;

$$\left[x^2 + \frac{b}{a}x + \left(\frac{b}{2a} \right)^2 \right] + \frac{c}{a} - \left(\frac{b}{2a} \right)^2 = 0$$

The terms in the square brackets are equal to $(x + b/2a)^2$

$$\therefore \quad \left(x + \frac{b}{2a} \right)^2 = \left(\frac{b}{2a} \right)^2 - \frac{c}{a}$$

$$= \frac{b^2}{4a^2} - \frac{c}{a}$$

77

$$= \frac{b^2 - 4ac}{4a^2}$$

$$\therefore \quad x + \frac{b}{2a} = \pm \sqrt{\left(\frac{b^2 - 4ac}{4a^2}\right)}$$

$$= \pm \frac{\sqrt{(b^2 - 4ac)}}{2a}$$

Plus and minus signs occur in front of the square root sign. This is because the square of a negative number is always a positive number, eg $(-2)^2 = 4$. Therefore the square root of a positive number can be either a positive or negative number, although its absolute value is always the same.

$$\therefore \quad x = \frac{-b}{2a} \pm \frac{\sqrt{(b^2 - 4ac)}}{2a}$$

$$= \frac{-b \pm \sqrt{(b^2 - 4ac)}}{2a}$$

The two roots of the quadratic equation are

$$x_1 = \frac{-b + \sqrt{(b^2 - 4ac)}}{2a} \quad \text{and} \quad x_2 = \frac{-b - \sqrt{(b^2 - 4ac)}}{2a}$$

Checking the previous example using this formula, we have

$$x^2 - 8x + 7 = 0$$

Here $a = 1$, $b = -8$, $c = 7$

$$\therefore \quad x = \frac{8 \pm \sqrt{(64 - 28)}}{2}$$

$$= 4 \pm \tfrac{1}{2}\sqrt{36} = 4 \pm 3$$

$$\therefore \quad x_1 = 1 \quad \text{and} \quad x_2 = 7$$

In the second example

$$x^2 - 8x + 16 = 0$$

$$x = \frac{8 \pm \sqrt{(64 - 64)}}{2} = 4$$

The equal roots are, therefore, obtained when the quantity under the square root sign is equal to zero; that is, when

$$b^2 = 4ac$$

In the third example

$$x^2 - 8x + 20 = 0$$

$$x = \frac{8 \pm \sqrt{(64 - 80)}}{2}$$

$$= 4 \pm \tfrac{1}{2}\sqrt{(-16)}$$

Here there is a negative quantity under the square root sign and this has no real value. There are, therefore, no real roots to this equation.

However, if we could obtain the square root of a negative number, we could obtain the roots of the equation. Notice that $(4)^2 = 16$ and $(-4)^2 = 16$ so that $\sqrt{(-16)}$ is not 4 or -4. We now introduce a new number, denoted by the letter i, which is defined by $i^2 = -1$ or $i = \sqrt{(-1)}$. Since there is no 'real' number which when squared gives -1, i is known as an *imaginary number*.

The properties of i are such that it can be used as an ordinary number bearing in mind that $i^2 = -1$, so that $i^3 = -i$ and $i^4 = 1$.

We can now write $-16 = 16(-1) = 16i^2$

Hence $$\sqrt{(-16)} = \sqrt{(16i^2)} = 4i$$

Therefore,

$$x = 4 \pm \tfrac{1}{2}\sqrt{(-16)} = 4 \pm \frac{4i}{2} = 4 \pm 2i$$

and the roots of $x^2 - 8x + 20$ are $x = 4 + 2i$ and $x = 4 - 2i$. Numbers such as these, consisting of a real and an imaginary part, are known as *complex numbers*. They are useful in several branches of economics, in particular in the solution of differential and difference equations arising from growth theory. These are discussed in Chapters 9 and 10.

The solutions of a quadratic equation can be summarised as follows:

If $ax^2 + bx + c = 0$

then $$x = \frac{-b \pm \sqrt{(b^2 - 4ac)}}{2a}$$

79

and if $b^2 > 4ac$ there are two real and distinct roots

if $b^2 = 4ac$ there are two real and coincident roots

if $b^2 < 4ac$ there are two complex roots

3.6 EXERCISES

1. Sketch the graphs of the average cost functions for the following total cost functions.

   (a) $TC = 50x$
   (b) $TC = 4 + x$
   (c) $TC = 63 + 6x + x^2$

   In each case, what is the limiting value of the average cost as $x \to \infty$? ($x =$ output).

2. A manufacturer has the following information about the costs of production on a machine:

Output (x)	5	10	15
Total costs	20	65	160

   (a) Determine the equation of the cost function, assuming it can be represented by a quadratic expression.
   (b) What is the cost of producing 20 units of output?
   (c) Sketch the graph of the cost function for $0 \leqslant x \leqslant 20$.
   (d) The revenue function is given by

   $$y_R = 10x + 15$$

   Show that there are two levels of output at which total revenue equals total costs, and illustrate this graphically.

3. The total cost function for an output of x units of a product is

   $$TC = 250 + 4x$$

   Draw the graph of the average cost function and show that the average cost approaches the value of 4 asymptotically.

4. Show that the equations

$$xy = 10$$

and $$3y = 32 - 2x$$

are satisfied for two values of x and sketch the graphs of the equations.

5. Show that the simultaneous equations

$$y = 20 + 3x + x^2$$

$$y = 5x + b$$

have the two real solutions when $b = 20$, two real coincident solutions when $b = 19$ and two complex solutions when $b = 18$. Sketch the graphs of these equations to illustrate these three cases.

3.7 SIMULTANEOUS QUADRATIC EQUATIONS

We have used linear equations to represent demand and supply functions but we are now able to extend the analysis a little further by considering quadratic equations. For example,

demand $\quad q_d = f(p) = p^2 - 8p + 15$

supply $\quad q_s = g(p) = 2p^2 + 3p - 3$

Graphical representation of these two equations on the same set of axes shows that there is a position where the curves intersect. This is the point at which supply and demand are equal (see Fig 3.9).

The equilibrium values of p and q could be obtained graphically but we can also solve the two quadratic equations simultaneously. For example,

$$q_d = p^2 - 8p + 15$$

$$q_s = 2p^2 + 3p - 3$$

and in equilibrium $q_d = q_s$

$$\therefore \quad p^2 - 8p + 15 = 2p^2 + 3p - 3$$

$$\therefore \quad p^2 + 11p - 18 = 0$$

$$\therefore \quad p = \frac{-11 \pm \sqrt{[11^2 + (4 \times 18)]}}{2} = \frac{-11 \pm \sqrt{193}}{2}$$

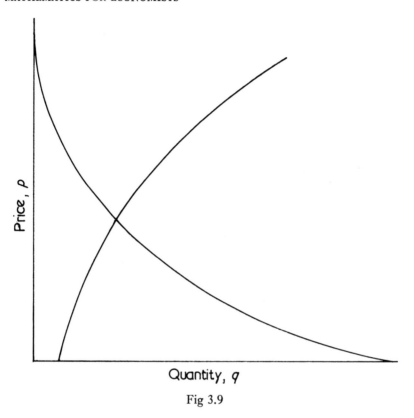

Fig 3.9

$\sqrt{193}$ can be found by reference to tables of square roots. The result is

$$p = \frac{-11 \pm 13.9}{2}$$

and the two roots are $p_1 = 2.9/2 = 1.45$ and $p_2 = -24.9/2 = -12.45$.

Mathematically the two solutions are both correct because the quadratic curves intersect in two places, but by the very nature of our problem price cannot be negative, and therefore we must take the positive root.

The equilibrium price will be 1.45 units and at this price

$$2.1 - 11.6 + 15 = 5.5 \text{ units are produced and sold.}$$

3.8 EXERCISES

1. Find the equilibrium price and quantity for the demand curve

$$q = 250 - 4p - p^2$$

and the supply curve

$$q = 2p^2 - 3p - 40$$

and sketch the curves for the range

$$p = 0 \quad \text{to} \quad p = 10$$

2. Determine the equilibrium price and quantity for the following demand and supply functions. Sketch the curves for the range $p = 0$ to $p = 8$

	Demand	*Supply*
(a)	$q - 25 + 3p = 0$	$q = 2p^2 - 40$
(b)	$p^2 + q^2 = 32$	$4 = 3p - 2q$
(c)	$pq = 6$	$q = 3(p - 1)$

3.9 AN ECONOMIC EXAMPLE

Quadratic equations often arise in economic theory from situations where the original data can be represented by a linear equation.

Let the demand function for a product be

$$q_D = f(p) = 1,000 - 10p$$

The total revenue, which is obtained by the sale of q items, is obtained by multiplying price by quantity sold.

$$\therefore \quad \text{Total revenue} = pq = p(1,000 - 10p)$$

$$= 1,000p - 10p^2$$

This is a quadratic equation and a graph of the function shows that the total revenue rises at first and then after reaching some maximum value starts to decline (Fig 3.10).

This form of graph is known as a *parabola*. It is symmetrical about the line parallel to the revenue axis through the point $p = 50$ as shown by the dotted line. The maximum revenue is therefore obtained when $p = 50$ and this is equal to

$$p(1,000 - 10p) = 50(1,000 - 500) = 25,000$$

83

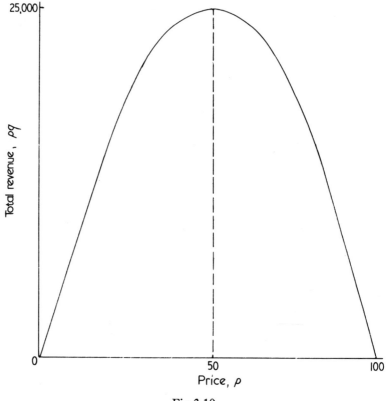

Fig 3.10

The quantity which will be sold at this price is given by

$$q_D = 1,000 - 10p = 1,000 - 500 = 500$$

The total revenue can be obtained as a function of quantity demanded from the equations

$$q_D = 1,000 - 10p \tag{1}$$

$$TR = pq = 1,000p - 10p^2 \tag{2}$$

from eq. (1) $10p = 1,000 - q_D$

$$\therefore \quad p = 100 - \tfrac{1}{10}q_D$$

Substituting this in eq. (2), we obtain

$$TR = 1,000(100 - \tfrac{1}{10}q_D) - 10(100 - \tfrac{1}{10}q_D)^2$$
$$= 100,000 - 100q_D - 100,000 + 200q_D - \tfrac{1}{10}q_D^2$$
$$= 100q_D - \tfrac{1}{10}q_D^2$$

This is again a quadratic equation which is a parabola and total revenue is equal to zero when

$$100q_D - \tfrac{1}{10}q_D^2 = 0$$
$$q_D(100 - \tfrac{1}{10}q_D) = 0$$

∴ either $$q_D = 0$$

or $$100 - \tfrac{1}{10}q_D = 0$$

that is, $$q_D = 1,000$$

The maximum revenue is equal to $500 \times 50 = 25,000$. The function can now be represented graphically as shown in Fig 3.11.

The company will be interested not so much in the total revenue but in the net revenue after the costs of production etc have been met. The line superimposed on the graph corresponds to the linear cost function.

$$TC = 5,000 + 15q_D$$

The net revenue for any quantity of output q_D is given by the vertical distance between these two functions at that quantity. Zero net revenue is obtained at the two breakeven points which are given by

$$TR = TC$$
$$100q_D - \tfrac{1}{10}q_D^2 = 5,000 + 15q_D$$

ie $$\tfrac{1}{10}q_D^2 - 85q_D + 5,000 = 0$$

$$q_D = \frac{85 \pm \sqrt{(85^2 - 4 \times \tfrac{1}{10} \times 500)}}{2 \times \tfrac{1}{10}} = \frac{85 \pm 72.28}{0.2}$$

That is approximately 64 units and 786 units. Between these two points is the quantity of output which yields the maximum net

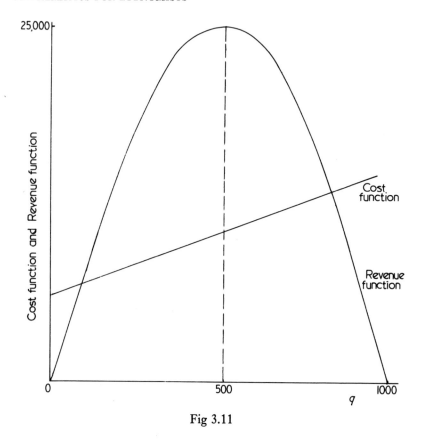

Fig 3.11

revenue and it could be obtained from the equation representing the net revenue function:

$$\text{Net revenue} = \text{total revenue} - \text{total cost}$$

$$= 100q_\text{D} - \tfrac{1}{10}q_\text{D}^2 - (5{,}000 + 15q_\text{D})$$

$$= 85q_\text{D} - \tfrac{1}{10}q_\text{D}^2 - 5{,}000$$

The maximum value of this function can be found graphically, but is obtained more simply using the differential calculus (see Chapter 5).

3.10 DISCONTINUOUS FUNCTIONS

So far we have looked at those functions for which the graph is a continuous line between the values considered. However, an equation may be valid only between certain limits. For example, the relationship between the total cost of manufacture and the number of units of output of a product may be represented by an equation such as the following:

$$TC = a + bx$$

where a is the fixed cost and b the variable cost of production which in this case is constant for all levels of output. But it is apparent that the volume of output cannot be increased indefinitely for a given amount

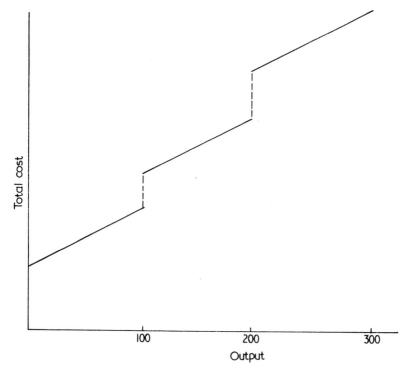

Fig 3.12

of fixed costs. For example extra investment in machinery will eventually be needed. We must, therefore, add the condition that the equation is valid only up to a certain quantity of output above which an increase in fixed costs must be incurred. By a stepwise construction we might arrive at the following set of equations.

$$TC = a_1 + bx \qquad 0 < x \leqslant 100$$

$$TC = a_2 + bx \qquad 100 < x \leqslant 200$$

$$TC = a_3 + bx \qquad 200 < x \leqslant 300$$

with the condition $a_1 < a_2 < a_3$.

This set of equations can be graphed on the one set of axes as in Fig 3.12. The graph in this case is not continuous between the limits $q = 0$ and $q = 300$. It consists of three separate sections and at the values $q = 100$ and $q = 200$ there are discontinuities. At these values a very small increase in one variable, quantity, requires a large increase in the other variable, total cost. In this particular example the discontinuity occurs when the value of the constant a in the equation is changed.

3.11 EXERCISES

1. A firm has a discontinuous cost function and the relationship between output (x) and total cost (TC) is given by

$$TC = 15 + 3x \qquad 0 \leqslant x \leqslant 30$$

$$TC = 40 + 3x \qquad 30 \leqslant x \leqslant 50$$

$$TC = 70 + 3x \qquad 50 \leqslant x \leqslant 100$$

Illustrate this graphically. What is the cost of raising output

(a) from 19 to 21 units?

(b) from 29 to 31 units?

(c) from 49 to 51 units?

CHAPTER FOUR

Series

4.1 INTRODUCTION

A *progression* is a number of terms which are arranged in a definite order. For example:

$$100, 101, 102, \quad 103, \quad 104, \ldots$$

$$1, \quad 10, 100, 1{,}000, 10{,}000, \ldots$$

$$0, 100, 100, \quad 200, \quad 400, \ldots$$

are all progressions. In the first example 100 is the *first term* or *initial term* and each subsequent term is obtained by adding 1 to the previous term. In the second example, 1 is the first term and each subsequent term is obtained by multiplying the previous term by 10. In the third example the first term is 0, the second term 100, and each subsequent term is obtained by adding together all the previous terms, so that $0 + 100 = 100$, $0 + 100 + 100 = 200$, $0 + 100 + 100 + 200 = 400$, and so on.

4.2 ARITHMETIC PROGRESSIONS

An *arithmetic progression* is one in which there is a constant difference between each term. The first example above is an arithmetic progression. The difference between any two (consecutive) terms is 1. Another example is the progression

$$100, 90, 80, 70, 60, \ldots$$

where 100 is the first term and the difference between any two terms is -10. In general, any arithmetic progression can be written as

$$a, a+d, a+2d, a+3d, \ldots$$

where a is the first term, and d is known as the *common difference*. For example, for the progression

$$100, 101, 102, 103, 104, \ldots$$

$$a = 100, \ d = 1$$

while for

$$100, 90, 80, 70, 60, \ldots$$

$$a = 100, \ d = -10.$$

Notice that in the general arithmetic progression the second term is $a + d$, the third term is $a + 2d$, the fourth term is $a + 3d$, and so the nth term is $a + (n-1)d$. For example for $a = 100$ and $d = 1$, the twentieth term is

$$a + (20 - 1)d = 100 + 19 = 119$$

and for $a = 100$, and $d = -10$, the tenth term is

$$a + (10 - 1)d = 100 + 9(-10) = 10$$

A useful application of arithmetic progressions is in the calculation of simple interest.

For example, suppose £100 is invested at 5 per cent per annum simple interest. Then the interest for the first year is £5 (from £100 × 0.05 = £5). Similarly, the interest for any subsequent year will be £5, and so the value of the investment is as shown in Table 4.1, since $a = 100$, $d = 5$.

TABLE 4.1

Year	1	2	3	4	...	n
Value at beginning of year	100	$100 + 5$ $= 105$	$105 + 5$ $= 110$	$110 + 5$ $= 115$		$100 + (n-1)5$

The value at the beginning of the tenth year (ie after 9 complete years) is $a + (10 - 1)d = 100 + 9(5) = £145$, and the value of the investment after 20 years (ie at the beginning of year 21) is

$$100 + (21 - 1)5 = £200$$

It is frequently useful to be able to obtain the sum of an arithmetic progression to give an *arithmetic series*. Returning to the general progression a, $a+d$, $a+2d$, $a+3d$, ..., $a+(n-1)d$, we see that the sum of the first n terms, $s(n)$, is

$$s(n) = a + (a+d) + (a+2d) + \cdots + [a+(n-1)d]$$
$$= n(a) + d + 2d + 3d + \cdots + (n-1)d$$

which can also be written (by reversing the order) as

$$s(n) = n(a) + (n-1)d + (n-2)d + \cdots + 3d + 2d + d$$

Adding these two expressions together gives

$$2s(n) = 2na + nd + nd + \cdots + nd$$

since such terms as $d + (n-1)d = d + nd - d = nd$ and

$$3d + (n-3)d = 3d + nd - 3d = nd$$

Hence,

$$2s(n) = 2na + (n-1)nd$$

as there are $(n-1)$ terms equal to nd.

$$\therefore \quad s(n) = \frac{n}{2}[2a + (n-1)d]$$

Alternatively, this can be written as

$$s(n) = n\frac{[a + \{a + (n-1)d\}]}{2}$$

= (number of terms) (average of first and last terms).

For example, for the progression 100, 101, 102, ... in which $a = 100$, $d = 1$, the sum of the first n terms is

$$s(n) = \frac{n}{2}[200 + (n-1)]$$

so that the sum of the first 3 terms is

$$s(3) = \tfrac{3}{2}[200 + (3-1)] = \frac{3(202)}{2} = 303$$

which can easily be verified. The sum of the first 20 terms is

$$s(20) = \frac{20}{2} [200 + (20 - 1)] = 2,190$$

which is less easily verified.

Further Example A man invests £100 per annum in bonds which pay 5 per cent per annum simple interest. What is the value of the investment at the beginning of the 11th year? This value is the sum of the first eleven terms of the arithmetic progression with $a = 100$, $d = 5$ and is given by

$$s(11) = \frac{11}{2} [200 + (11 - 1)5] = £1,375$$

This is verified by looking at the value of the investment for the first few years (Table 4.2). Thus, at the beginning of the 4th year the value is $100 + 105 + 110 + 115$, which is the sum of the first 4 terms of the arithmetic progression with $a = 100$, $d = 5$. Hence the value of the investment at the beginning of the 11th year (ie when $n = 11$) is

$$s(11) = \frac{11}{2} [200 + (11 - 1)5] = £1,375.$$

TABLE 4.2

Year	Value at beginning of year
1	100
2	$100 + (100 + 5)$
3	$100 + (100 + 5) + (100 + 10)$
4	$100 + (100 + 5) + (100 + 10) + (100 + 15)$

4.3 EXERCISES

1. Give the tenth term of each of the following progressions and also determine the sum of the first fifteen terms.

(a) 1, 3, 5, 7, . . .

(b) 500, 550, 600, 650, . . .

(c) 60, 30, 0, -30, . . .

(d) 100, 98, 96, 94, . . .

2. What is the value of an investment of £200 after 5 years if simple interest is paid at 10 per cent per annum?

3. A man receives a salary of £850 per annum which increases annually by £40. What salary will he receive in 6 years time? What will be his total income during those 6 years?

4. A government issues savings bonds which sell for £10 and are worth £20 after 15 years. What is the implied rate of simple interest?

4.4 GEOMETRIC PROGRESSIONS

A geometric progression is one in which there is a constant ratio between any two consecutive terms. For example in the progression

$$1, 10, 100, 1,000, 10,000, . . .$$

each term is 10 times the previous term so that $100 = 10(10)$ and $10,000 = 10(1,000)$. In the progression 128, 64, 32, 16, 8, . . . each term is 0.5 times the previous term, eg $64 = 0.5(128)$, and $32 = 0.5(64)$. In general, any geometric progression can be written as

$$a, ar, ar^2, ar^3, . . .$$

where a is the first term and r is known as the *common ratio*. In the above examples, $a = 1$, $r = 10$ and $a = 128$, $r = 0.5$. The nth term in a geometric progression is ar^{n-1}, so that for the progression with $a = 1$, $r = 10$, the sixth term is $1(10^5) = 100,000$, while for the progression with $a = 128$, $r = 0.5$, the fifth term is $128(0.5^4) = 8$.

One of the applications of geometric progressions is in the calculation of compound interest. Here the sum on which interest is paid includes the interest which has been earned in previous years.

For example, if £100 is invested at 5 per cent per annum compound interest, then after 1 year the interest earned is £5 (from 100×0.05) and the capital invested for the second year is £105. The interest earned by this capital is not £5 but £105 $\times 0.05 = $ £5.25. The capital invested for the third year is £105 + £5.25 = £110.25, and the interest earned is £110.25 $\times 0.05 = $ £5.5125. This is shown in Table 4.3.

TABLE 4.3

Beginning of year	Capital	Interest during year
1	100	5
2	105	5.25
3	110.25	5.5125
4	115.7625	5.7881

This can be expressed in general terms, as shown in Table 4.4 for an investment of a and an interest rate of i per cent. This shows that the progression for the capital is a, $a(1+i)$, $a(1+i)^2$, $a(1+i)^3, \ldots$, which is of a geometric form with common ratio $r = 1+i$.

TABLE 4.4

Beginning of year	Capital	Interest during year
1	a	ai
2	$a + ai = a(1+i)$	$a(1+i)i$
3	$a(1+i) + a(1+i)i = a(1+i)^2$	$a(1+i)^2 i$
4	$a(1+i)^2 + a(1+i)^2 i = a(1+i)^3$	$a(1+i)^3 i$

For example, if $i = 5$ per cent $= 0.05$, then $r = 1.05$ and so the nth term in the progression if $a = 100$ is $100(1.05)^{n-1}$. The value of the investment after 8 years is $100(1.05)^{9-1} = £147.75$ (note that $n = 9$ since we want the ninth term in the progression). The evaluation of such quantities as 1.05^8 is best done using a table of logarithms. A short section on the use of logarithms is included at the end of this chapter (Section 4.14).

The value of a *geometric series*, which is the sum of a geometric progression, is easily obtained by considering the general progression

$$a, \ ar, \ ar^2, \ ar^3, \ \ldots, \ ar^{n-1}$$

The sum of the first n terms,

$$S(n) = a + ar + ar^2 + ar^3 + \cdots + ar^{n-1}$$

also $\qquad r \cdot S(n) = ar + ar^2 + ar^3 + \cdots + ar^{n-1} + ar^n$

Subtracting gives $S(n) - rS(n) = a - ar^n$; that is

$$S(n)(1-r) = a(1-r^n) \qquad \text{or} \qquad S(n) = \frac{a(1-r^n)}{1-r}$$

For example, for the progression with $a = 1$, $r = 10$,

$$S(n) = \frac{1(1-10^n)}{1-10}$$

so that the sum of the first 5 terms is

$$S(5) = \frac{1(1-10^5)}{1-10} = \frac{10^5 - 1}{9} = 11{,}111$$

and the sum of the first 8 terms is

$$S(8) = \frac{1(1-10^8)}{1-10} = \frac{10^8 - 1}{9} = 11{,}111{,}111$$

A special case arises when $-1 < r < 1$.

Then $\qquad\qquad\qquad S(n) = \dfrac{a}{1-r} - \dfrac{ar^n}{1-r}$

and as $n \to \infty$, $r^n \to 0$, and the second term will approach zero.

The sum of all the terms in the series or, as it is generally referred to, the sum to infinity is given by

$$S(\infty) = \frac{a}{1-r}$$

For example, if $a = 128$, $r = 0.5$

$$S(\infty) = \frac{128}{1-0.5} = \frac{128}{0.5} = 256$$

This means that adding together the successive terms of the progression 128, 64, 32, 16, 8, 4, 2, 1, 0.5, ... produces a sum which becomes closer and closer to, but never becomes greater than, 256.

It is important to realise that r must satisfy $-1 < r < 1$ for the sum to infinity to be finite. For example, for the series $a = 1$, $r = 10$,

$$S(\infty) = \frac{1}{1-10} - \frac{1.10^\infty}{1-10} = -\frac{1}{9} + \frac{10^\infty}{9}$$

which is obviously infinity.

4.5 EXERCISES

1. State the sixth term of each of the following progressions and determine the sum of the first 10 terms. Evaluate the sum to infinity if it is finite.

   (a) 10, 30, 90, 270, . . .

   (b) 81, 27, 9, 3, . . .

   (c) 2, −4, 8, −16, . . .

   (d) −1,024, 512, −256, 128, . . .

   (e) 1, 0.1, 0.01, 0.001, . . .

2. What is the value of an investment of £300 at the end of 15 years if compound interest is paid at

   (a) 5 per cent per annum?   (b) 10 per cent per annum?

3. An insurance policy costs £150 per annum for 25 years. What is the final value of the policy if compound interest is earned at a rate of 6 per cent per annum?

4. Government savings bonds are sold for £4 and are worth £5 after 4 years. What is the implied rate of compound interest?

4.6 DISCOUNTING

If money is able to earn interest its absolute value in future years will be greater than its current or present value. Conversely money which is to be spent in future years has a present value which is less than its absolute value.

For example, if we have £100 now and we invest this at 6 per cent per annum compound interest, then we will have £106 at the end of one year. If, therefore, we are required to spend £106 in one year's time we would only need to have available at the present time £100 which we could invest at 6 per cent. The present value of a capital sum of £106 which is required in one year's time is therefore equal to £100 when the interest rate is 6 per cent per annum.

A sum of £100 required in one year's time then has a present value given by

$$PV = \frac{100}{106} \times 100 = \frac{100}{1.06} = \frac{100}{(1+6/100)} = £94.3$$

This method of calculating the present value of future sums of money, whether payments or receipts, is known as *discounting*. It can be used to cover any time span as follows:

If future requirements for capital are known with their appropriate time pattern then

$$(PV)_C = a_0 + \frac{a_1}{(1+i)} + \frac{a_2}{(1+i)^2} + \cdots + \frac{a_n}{(1+i)^n}$$

where $(PV)_C$ = present value of capital requirements

a_0 is the immediate requirement

a_1 is the requirement in the first year and discounted as though it appeared at the end of the first year.

a_2 is the requirement in the second year
$\vdots$
a_n is the requirement in the nth year.

If these capital requirements or cash outflows are for a given project then it must be assumed that the project will produce a series of receipts, or cash inflows, again over a period of time. These can be discounted using the same compound interest rate.

$$(PV)_R = b_0 + \frac{b_1}{(1+i)} + \frac{b_2}{(1+i)^2} + \cdots + \frac{b_n}{(1+i)^n}$$

where $(PV)_R$ = present value of receipts

b_0 is the immediate return

b_1 is the return in the first year and discounted as though it appeared at the end of the first year

b_2 is the return in the second year

$\vdots$

b_n is the return in the nth year.

For the project to be profitable on a strictly financial basis we must have the condition that

$$(PV)_R > (PV)_C$$

that is, the present value of the receipts from the project must be greater than the present value of the payments.

The above is known as the *present-value method* for assessing investment projects and the *net present value (NPV)* is defined as the difference between the discounted values of receipts and payments.

$$NPV = (PV)_R - (PV)_C$$

The value used for i in the discounting process obviously has a considerable influence on the net present value. It is known as the cost of capital and its value is often difficult to establish accurately in many practical situations. Because of this, preference is often given to a method which allows the cost of capital to be a variable and determines that value which equates the present value of receipts to the present value of payments. That is,

$$(PV)_R = (PV)_C$$

or $$a_0 + \frac{a_1}{(1+i)} + \cdots + \frac{a_n}{(1+i)^n} = b_0 + \frac{b_1}{(1+i)} + \cdots + \frac{b_n}{(1+i)^n}$$

From this equation the values of i can be determined. Since i is a rate of interest it must be positive, and normally we expect only one positive value to result. This value is known as the *internal rate of return (IRR)* and the higher this is the more profitable the project will be.

Both of the above methods, ie *NPV* and *IRR*, require that each year's payments and receipts are discounted by the appropriate factor which is determined by the time period in which these payments and receipts arise. Tables containing factors which simplify the arithmetic are available and a selection of these is reproduced in Table 4.5. It is also possible to carry out the complete exercise to produce either *NPV* or i on a computer.

TABLE 4.5

Table of Discount Factors

This table shows the present value of 1 discounted for different numbers of years and at different rates of discount

Rate of discount (per cent) Year	8	10	12	14	16
0	1.000	1.000	1.000	1.000	1.000
1	0.926	0.909	0.893	0.877	0.862
2	0.857	0.826	0.797	0.769	0.743
3	0.794	0.751	0.712	0.675	0.641
4	0.735	0.683	0.636	0.592	0.552
5	0.681	0.621	0.567	0.519	0.476
6	0.630	0.564	0.507	0.456	0.410
7	0.583	0.513	0.452	0.400	0.354
8	0.540	0.467	0.404	0.351	0.305
9	0.500	0.424	0.361	0.308	0.263
10	0.463	0.386	0.322	0.270	0.227
15	0.315	0.239	0.183	0.140	0.108
20	0.215	0.149	0.104	0.073	0.051

4.7 EXERCISES

1. (a) If £300 is invested for 10 years and is then worth £800, what is the implied rate of compound interest?
   (b) What is the present value of the £800 at 16 per cent per annum compound interest?
2. What is the present value of £1,000 payable after 10 years if the rate of discounting is
   (a) 14 per cent?   (b) 8 per cent?
3. (a) How much should be invested at 10 per cent per annum compound interest to give £250 after 5 years?
   (b) What is the value of this investment after 3 years?
4. Two projects are available to a company and the estimated returns are:

End of year	1	2	3
Project A	100	200	300
Project B	150	300	100

Which project has the greater present value if the discounting rate is 10 per cent?

5. What is the internal rate of return from a project which has the following costs and receipts?

End of year	1	2	3
Costs	120	120	100
Receipts	100	110	160

4.8   ANNUITIES AND SINKING FUNDS

An *annuity* is a constant annual income which can be bought for cash. For example, an annual income of £10 for 20 years may be purchased for £120. The present value of an annuity of £A payable for n years is given by

$$p = \frac{A}{1+i} + \frac{A}{(1+i)^2} + \frac{A}{(1+i)^3} + \cdots + \frac{A}{(1+i)^n}$$

$$= \frac{A}{1+i}\left[1 + \frac{1}{1+i} + \frac{1}{(1+i)^2} + \cdots + \frac{1}{(1+i)^{n-1}}\right]$$

The term in the brackets is the sum of a geometric series with $a = 1$ and $r = 1/(1+i)$ and so

$$p = \frac{A}{1+i}\left[\frac{1-r^n}{1-r}\right] = \frac{A}{i}\left[1 - \frac{1}{(1+i)^n}\right]$$

For example, the present value of an annuity of £10 per annum for 20 years if the rate of interest is 7 per cent is

$$p = \frac{10}{0.07}\left[1 - \frac{1}{(1.07)^{20}}\right] = £105.94$$

A special case occurs when the annuity continues indefinitely, since

$$p = \frac{A}{1+i} + \frac{A}{(1+i)^2} + \frac{A}{(1+i)^3} + \cdots = \frac{A/(1+i)}{1-1/(1+i)} = \frac{A}{i}$$

For example, an annuity of £20 per annum for ever has a present value of

$$p = \frac{20}{0.05} = £400$$

if the rate of interest is 5 per cent per annum.

A *sinking fund* is a fund set up to meet some financial commitment, to which a constant sum is added each year. For example, how much should be invested each year if the rate of interest is 5 per cent per annum to give a capital sum of £600 after 10 years?

Let A be the amount invested. After 1 year, the value of the investment will be $A + A(1+i) = A + A(1.05)$. After 2 years, the value will be $A + A(1.05) + A(1.05)^2$ and after 10 years the value will be

$$A + A(1.05) + A(1.05)^2 + \cdots + A(1.05)^9$$

This has to be equal to £600.

$$\therefore \quad 600 = A(1 + 1.05 + 1.05^2 + \cdots + 1.05^9)$$
$$= \frac{A[1 - (1.05)^{10}]}{1 - 1.05}$$

since the expression in brackets is a geometric series.

$$\therefore \quad A = \frac{600(1 - 1.05)}{1 - (1.05)^{10}} = £47.7$$

That is, if a sum of £47.7 is invested each year it will accumulate to £600 in 10 years at an interest rate of 5 per cent per year.

4.9 EXERCISES

1. What is the present value of an annuity of £100 per annum which is to be paid 20 times, commencing one year from now, if the interest rate is 8 per cent?

2. A chief rent is an annual payment for land which is paid indefinitely. What is the present value of a chief rent of £25 if the rate of interest is 4 per cent?

3. A company buys a machine for £2,000 and estimates that its life will be 15 years. How much should be paid annually into a

sinking fund to buy a replacement for the machine in 15 years time if the replacement will cost £2,500 and the rate of interest is 9 per cent?

4. An ex-pupil decides to donate to his school a sum of money to provide an annual prize of £6 for the next ten years.

   (a) How much should he donate if the rate of interest is 6 per cent?

   (b) How much should he donate if the prize is to be paid indefinitely?

4.10   INTEREST PAID CONTINUOUSLY

Suppose that a man has £1 which he invests at a rate of interest of 100 per cent a year compound. The value of his investment at the end of a year is £1 + £1 = £2, since he gains £1 interest on his capital. However, let us now suppose that the interest is paid twice a year, after the first six months and at the end of the year. After the first six months the interest due is £0.5 because the £1 has been invested for half a year. But his capital is now £1.5 and so the interest due for the second six months of the year is $0.5 \times 1.5 = 0.75$, giving a total value to the investment of £2.25 at the end of the year. In the same way, if we assume that the interest is paid at the end of each quarter, the capital value at the end of the first quarter is £(1 + 0.25), at the end of the second quarter £1.25(1 + 0.25), at the end of the third quarter £1.25^2 (1 + 0.25) and at the end of the year £1.25^3(1 + 0.25) $= 1.25^4 = 2.44$.

In each of these cases the value of £1 at the end of the year is given by

$$(1 + 1/n)^n$$

where n is the number of times the interest is paid during the year.

That is,

$$n = 1, \quad \left(1 + \frac{1}{1}\right)^1 = £2$$

$$n = 2, \quad \left(1 + \frac{1}{2}\right)^2 = 1.5^2 = £2.25$$

$$n = 4, \quad \left(1 + \frac{1}{4}\right)^4 = 1.25^4 = £2.44$$

We can draw up a table of values of $(1+1/n)^n$ for different values of n.

n	1	2	4	8	10	100	1,000	10,000
$(1+1/n)^n$	2.00	2.25	2.44	2.56	2.59	2.70	2.717	2.718

As the value of n increases the value of the investment becomes larger, but never exceeds £2.719. The limit to which the value tends as n tends to infinity (ie becomes indefinitely large) is defined as e,

that is, $$e = \lim_{n \to \infty} \left(1 + \frac{1}{n}\right)^n$$

and e is approximately 2.7183. This is known as an *irrational* number. Its value cannot be determined exactly no matter how many decimal places of working are used. Also, it cannot be represented by the ratio of two whole numbers.

By allowing n to approach infinity interest is being added to the investment more and more frequently and can be regarded as being added continuously. While this may not be realistic in the case of a monetary investment, the concept is very useful for other variables which tend to vary continuously, such as population, technical knowledge and education.

If the man invests £1 at x per cent per annum compound interest and the interest is paid n times in the year, the value of the investment at the end of the year is given by

$$\left(1 + \frac{x}{n}\right)^n$$

and the limit of this as n approaches infinity can be shown to be e^x.

4.11 BINOMIAL THEOREM

An expression of the form $(a+x)^n$ can generally be expanded as a series by applying the normal laws of algebra; for example,

$$(a+x)^2 = (a+x)(a+x) = a^2 + 2ax + x^2$$
$$(a+x)^3 = (a+x)(a+x)^2 = a^3 + 3a^2x + 3ax^2 + x^3$$

103

The algebra becomes a little tedious as the power of the function is increased but it is possible to obtain the expansion in the general case from the *binomial theorem*. This can be stated as follows:

$$(a+x)^n = {}^nC_0 a^n + {}^nC_1 a^{n-1}x + {}^nC_2 a^{n-2}x^2 + \cdots + {}^nC_r a^{n-r}x^r + \cdots + {}^nC_n x^n$$

where
$${}^nC_0 = 1$$

$${}^nC_1 = n$$

$${}^nC_2 = \frac{n(n-1)}{1 \times 2} = \frac{n(n-1)}{2!}$$

$${}^nC_3 = \frac{n(n-1)(n-2)}{1 \times 2 \times 3} = \frac{n(n-1)(n-2)}{3!}$$

and in general

$${}^nC_r = \frac{n(n-1)(n-2)\ldots(n-r+1)}{r!} = \frac{n!}{r!(n-r)!}$$

$r!$ is known as *factorial r* or *r* factorial and is the product of all the integer values from 1 to r. For example, $4! = 1 \times 2 \times 3 \times 4$. The exception to this is $0!$, which is defined to be equal to 1, so that

$${}^nC_n = \frac{n!}{n!(n-n)!} = \frac{n!}{n!0!} = \frac{n!}{n!} = 1$$

The coefficients of the terms in a binomial expansion form a symmetrical pattern, as can be seen in Fig 4.1, known as Pascal's triangle. Each number is the sum of the two numbers on the row above which are closest to it. Thus, the coefficient 10 is the sum of the two numbers 4 and 6 above it.

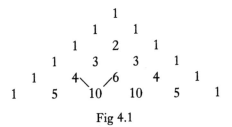

Fig 4.1

The third line of the triangle gives us the coefficients of $(a+x)^2$:

$$(a+x)^2 = \underline{1}a^2 + \underline{2}ax + \underline{1}x^2$$

The fourth line gives the coefficients of $(a+x)^3$:

$$(a+x)^3 = \underline{1}a^3 + \underline{3}a^2x + \underline{3}ax^2 + \underline{1}x^3$$

The fifth line gives the coefficients of $(a+x)^4$:

$$(a+x)^4 = \underline{1}a^4 + \underline{4}a^3x + \underline{6}a^2x^2 + \underline{4}ax^3 + \underline{1}x^4$$

and so on.

The expansion can be divided throughout by the constant a with the result

$$(a+x)^n = a^n \left(1 + \frac{x}{a}\right)^n$$

$$= a^n \left[1 + n\left(\frac{x}{a}\right) + \frac{n(n-1)}{2!}\left(\frac{x}{a}\right)^2 + \cdots \right.$$

$$\left. + \frac{n(n-1)\cdots(n-r+1)}{r!}\left(\frac{x}{a}\right)^r + \cdots + \left(\frac{x}{a}\right)^n\right]$$

The expansion in square brackets is particularly interesting when $n = -1$ since

$$\left(1 + \frac{x}{a}\right)^{-1} = 1 + (-1)\left(\frac{x}{a}\right) + \frac{(-1)(-2)}{1.2}\left(\frac{x}{a}\right)^2$$

$$+ \cdots + \frac{(-1)(-2)\cdots(-r)}{1.2.3\cdots r}\left(\frac{x}{a}\right)^r + \cdots$$

This series contains an infinite number of terms and the coefficient of every term is equal to unity, the signs of these coefficients being alternately positive and negative. That is,

$$\left(1 + \frac{x}{a}\right)^{-1} = 1 - \left(\frac{x}{a}\right) + \left(\frac{x}{a}\right)^2 - \left(\frac{x}{a}\right)^3 + \cdots + (-1)^r\left(\frac{x}{a}\right)^r + \cdots$$

When the numerical value of x/a is less than 1 the terms become smaller and smaller as x/a is raised to successively higher powers and eventually tend to the value zero. That is,

105

$$(-1)^r \left(\frac{x}{a}\right)^r \to 0 \quad \text{as} \quad r \to \infty \quad \text{for} \quad \left|\frac{x}{a}\right| < 1$$

In this particular case the sum of the series tends or converges to a finite value as more and more terms are considered. Note the condition $|x/a| < 1$. This means the numerical value, or *modulus*, of x/a must be less than 1. In general it can be shown that a series converges if successive terms tend to zero and also alternate in sign.

It can quite easily be seen that there is an infinite series which converges to a finite value for the function $(1 - x/a)^{-1}$ when the numerical value of x/a is less than one, since this is a geometric series with a common ratio between -1 and $+1$:

$$\left(1 - \frac{x}{a}\right)^{-1} = 1 + (-1)\left(-\frac{x}{a}\right) + \frac{(-1)(-2)}{1.2}\left(-\frac{x}{a}\right)^2 + \cdots$$

$$= 1 + \left(\frac{x}{a}\right) + \left(\frac{x}{a}\right)^2 + \cdots + \left(\frac{x}{a}\right)^r + \cdots$$

In this case all the terms are positive.

4.12 THE EXPONENTIAL SERIES

The binomial theorem can be used to evaluate e. We know that e is defined by

$$e = \lim_{n \to \infty} \left(1 + \frac{1}{n}\right)^n$$

and so using the binomial theorem to expand $(1 + 1/n)^n$, ie with $a = 1$, and $x = 1/n$,

$$\left(1 + \frac{1}{n}\right)^n = {}^nC_0 1 + {}^nC_1\left(\frac{1}{n}\right) + {}^nC_2\left(\frac{1}{n}\right)^2 + {}^nC_3\left(\frac{1}{n}\right)^3 + \cdots + {}^nC_n\left(\frac{1}{n}\right)^n$$

$$= 1 + \frac{n}{n} + \frac{n(n-1)}{2!}\frac{1}{n^2} + \frac{n(n-1)(n-2)}{3!}\frac{1}{n^3} + \cdots + \frac{1}{n^n}$$

$$= 1 + 1 + \left(\frac{n}{n}\right)\left(\frac{n-1}{n}\right) \cdot \frac{1}{2!}$$

$$+ \left(\frac{n}{n}\right)\left(\frac{n-1}{n}\right)\left(\frac{n-2}{n}\right)\frac{1}{3!} + \cdots + \frac{1}{n^n}$$

Allowing n to tend to infinity, all the ratios such as $(n-1)/n$, $(n-2)/n, \ldots$ tend to the value 1 and hence

$$e = \lim_{n \to \infty} \left(1 + \frac{1}{n}\right)^n = 1 + 1 + \frac{1}{2!} + \frac{1}{3!} + \frac{1}{4!} + \cdots$$

where the rth term in the expansion is $1/(r-1)!$.

This expansion, known as the *series expansion of e* can be used to evaluate e to any number of decimal places. For example to obtain e correct to 4 decimal places we continue summing the terms of the expansion until a term is less than 0.0001, as in Table 4.6.

TABLE 4.6

Number of terms summed	Last term	Contribution to e
1	1	1.00000
2	1	1.00000
3	$\frac{1}{2!}$	0.50000
4	$\frac{1}{3!}$	0.16667
5	$\frac{1}{4!}$	0.04167
6	$\frac{1}{5!}$	0.00833
7	$\frac{1}{6!}$	0.00139
8	$\frac{1}{7!}$	0.00020
9	$\frac{1}{8!}$	0.00002
		Total 2.81828

The approximate value of e from the first 9 terms of the expansion is 2.81828.

107

Similarly, e^x can be expanded as a power series to give

$$e^x = 1 + x + \frac{x^2}{2!} + \frac{x^3}{3!} + \frac{x^4}{4!} + \cdots + \frac{x^r}{r!} + \cdots$$

This expansion is known as the *exponential series*. It can be used to evaluate e^x to any degree of accuracy. Values of e^x are included in standard books of logarithm tables.

4.13 EXERCISES

1. £150 can be invested in one of three ways. It can receive 6 per cent per annum simple interest, 5 per cent per annum compound interest, or 4 per cent per annum compound interest with the interest compounded twice a year. Which is the most profitable investment over a period of
   (a) five years? (b) ten years?
2. What is the value of £50 after four years if it is invested at 6 per cent per annum compound interest with the interest compounded three times a year?
3. Use the binomial theorem to expand

   (a) $(1+x)^5$ (b) $(2-x)^3$

   (c) $\left(3+\dfrac{1}{x}\right)^4$ (d) $(2-x^2)^6$

4. Use the exponential series to obtain approximately the values of

   (a) $e^{0.1}$ (b) $e^{0.5}$
   (c) e^2 and (d) e^{-1}

   and compare your results with the values given in standard tables of e.

4.14 LOGARITHMS

We know that
$$100 \times 1,000 = 100,000$$

and since $100 = 10^2$, $1,000 = 10^3$ and $100,000 = 10^5$ then

$$10^2 \times 10^3 = 10^5.$$

That is, the product of 100 and 1,000 can be obtained by writing these numbers as powers of 10 and then adding these powers together. These powers are known as *logarithms* and are referred to as *logarithms to the base 10* because they are powers of 10.

Therefore,
$$\log_{10} 100 = 2$$
$$\log_{10} 1,000 = 3$$
$$\log_{10} 100,000 = 5$$

or, more simply
$$\log 100 = 2$$
$$\log 1,000 = 3$$
$$\log 100,000 = 5$$

Now $10 = 10^1$ and so $\log 10 = 1$

$0.1 = 10^{-1}$ and so $\log 0.1 = -1$

$1 = 10^0$ since $10 \times 0.1 = 1 = 10^1 \times 10^{-1} = 10^0$

and so $\log 1 = 0$.

Logarithms of any number in the range 1–10 can be found by using logarithm tables.

For example,
$$\log 2 = 0.3010$$
$$\log 5 = 0.6990$$
$$\log 7.5 = 0.8751$$
$$\log 3.001 = 0.4772$$
$$\log 1.055 = 0.0232$$

Logarithms of numbers outside the range 1–10 can be found by expressing the number as the product of a number within this range and ten raised to some power. For example, $500 = 5 \times 10^2$.

Now since $\log 5 = 0.6990$, $5 = 10^{0.6990}$

that is, $\log 500 = 2.6990$,

or $\log 500 = \log 5 + \log 100 = 0.6990 + 2 = 2.6990$

Similarly $0.003001 = 3.001 \times 10^{-3}$ and so $\log (0.003001) = \log (3.001) + \log (10^{-3}) = 0.4772 + (-3) = \bar{3}.4772$.

109

Notice that we write this as $\bar{3}.4772$ and not as $-3+.4772=-2.5228$. The reason for this is that it is useful for manipulation if the part of the logarithm after the decimal point is always positive.

The general rules governing logarithms are:

1. $\log ab = \log a + \log b$

   We have already used this rule above. For example, $a=5$, $b=100$, $ab=500$.

   $$\log 500 = \log 5 + \log 100 = 2.6990.$$

2. $\log \left(\dfrac{a}{b}\right) = \log a - \log b$

   for example, $a=5$, $b=2$,

   $\log (\frac{5}{2}) = \log 5 - \log 2 = 0.6990 - 0.3010 = 0.3980$

   that is, $\log 2.5 = 0.3980$

3. $\log a^n = n \log a$

   This follows directly from Rule 1.

   For example, $\log (a^2) = \log (a \cdot a) = \log a + \log a = 2 \log a$

   Similarly, $\log (5^3) = 3 \log 5 = 3(0.6990) = 2.0970$

These rules can be used to reduce the arithmetic required in our calculations.

Example 1 Evaluate $450/1.05^6$

By Rule 1, $\log 450 = \log (100 \times 4.5) = \log 100 + \log 4.5$

$$= 2 + 0.6532 = 2.6532$$

By Rule 3, $\log (1.05^6) = 6 \log (1.05) = 6(0.0212)$

$$= 0.1272$$

By Rule 2, $\log \left(\dfrac{450}{1.05^6}\right) = \log 450 - \log (1.05^6)$

$$= 2.6532 - 0.1272 = 2.5260$$

We now need to know the number for which 2.5260 is the logarithm. We can find this either by looking in the body of the logarithm tables for 0.5260, or by using tables of antilogarithms which give the required number directly.

$$\text{antilog } 0.5260 = 3.357$$

and therefore, antilog $2.5260 = 10^2 \times 3.357 = 335.7$; that is,

$$\frac{450}{1.05^6} = 335.7$$

Example 2 Evaluate $50(1.1^{10})/3250$.

We first of all evaluate 1.1^{10}, then $50(1.1^{10})$ and finally divide by 3250.

$$\log (1.1^{10}) = 10 \log 1.1 = 10(0.0414) = 0.4140$$

$$\log 50 = \log (10 \times 5) = \log 10 + \log 5 = 1.6990$$

$$\therefore \quad \log [(50(1.1^{10})] = \log (1.1^{10}) + \log 50 = 0.4140 + 1.6990$$

$$= 2.1130$$

$$\log 3250 = \log (1{,}000 \times 3.25) = 3.5119$$

$$\therefore \quad \log \frac{50(1.1^{10})}{3250} = \log [(50(1.1^{10})] - \log (3250)$$

$$= 2.1130 - 3.5119$$

$$= -1.3989 = -2 + 0.6011 = \bar{2}.6011$$

Hence

$$\left(\frac{50(1.1^{10})}{3250}\right) = \text{antilog } \bar{2}.6011$$

$$= 10^{-2} \text{ antilog } (0.6011)$$

$$= 10^{-2} (3.991)$$

$$= 0.03991$$

Throughout this section all the logarithms used have been to the base 10. This is generally convenient in numerical work. However, in the next chapter we are going to see that for most theoretical work it is more convenient to use logarithms to the base e. Such logarithms are known as natural, hyperbolic or naperian logarithms.

CHAPTER FIVE

Differential Calculus

5.1 INTRODUCTION

In Chapters 1 and 3 we saw that an equation relating two variables could be represented by a graph. For example, if y is the total cost of production and x the quantity produced then the equation

$$y = 100 + 3x$$

can be represented by a straight line, whilst

$$y = 100 + 2x + \tfrac{1}{10}x^2$$

can be represented by a curve. The graphs of these equations allow the cost to be determined for permissible levels of production.

But the equations also provide additional information which is very important. They tell us the change in the costs which must be incurred for any given change in production level. This can be determined directly from the graphs by considering a change of x from x_1 to x_2.

In the case of the linear cost function the cost changes from y_1 to y_2 (Fig 5.1). The average cost of each extra unit of production between the values x_1 and x_2 is given by the quotient

$$\frac{y_2 - y_1}{x_2 - x_1}$$

This is the slope of the line and it does not depend upon the value of x_1 or x_2, because the slope is constant for all values of x for a linear function.

This is not so for the quadratic function as can be seen from Fig 5.2. It is clear that with this curve the slope between P_1 and P_2 depends

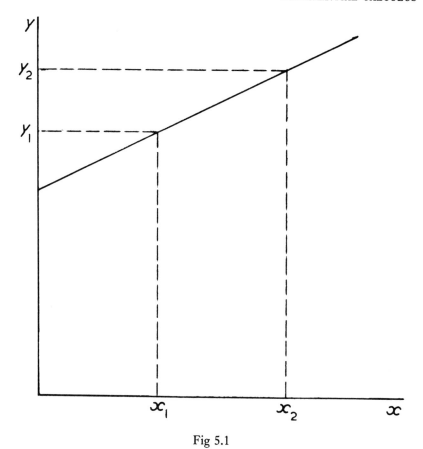

Fig 5.1

on where P_1 and P_2 are, and that even if P_1 is held constant, the slope will depend on the position of P_2. If the cost changes from y_3 to y_4 when the quantity produced changes from x_1 to x_2, then the average cost of each extra unit is

$$\frac{y_4 - y_3}{x_2 - x_1}$$

This is the slope of the straight line joining P_1 and P_2 and is not that of the curve. We therefore define the *slope at a point* on a curve as

113

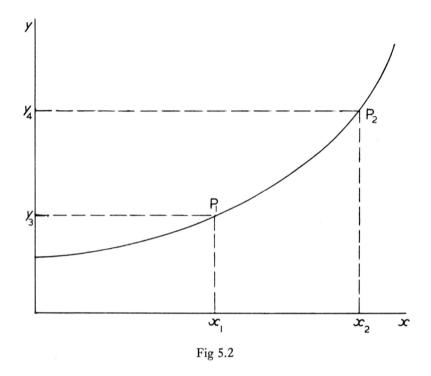

Fig 5.2

being a tangent to the curve at that point, where a *tangent* is a line which touches but does not cross the curve. The tangent at P_1 is shown in Fig 5.3, which is an enlargement of Fig 5.2 with P_1 and P_2 close together.

To indicate that only small changes in x and y are being represented we use the symbols δx and δy, which are *not* δ times x and δ times y. The slope of the straight line joining P_1 and P_2 is

$$\frac{\delta y}{\delta x}$$

But this is not the slope of the curve at P_1 (the tangent at P_1). However, if δx becomes smaller then the slope of the straight line $P_1 P_2$ becomes very close to the slope of the tangent at P_1.

114

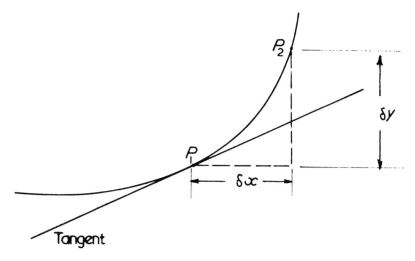

Fig 5.3

As $\delta x \to 0$, we define limit $(\delta y/\delta x)$ as the *derivative* of y with respect to x, and we denote it by the symbol dy/dx. That is,

$$\frac{dy}{dx} = \lim_{\delta x \to 0} \left(\frac{\delta y}{\delta x} \right)$$

The process of finding dy/dx is known as *differentiation* and dy/dx is sometimes called the *differential coefficient* of y with respect to x, rather than the derivative. We now discuss some methods of differentiation which will enable us to determine the slope of a function at any point.

In the case of the linear equation $y = 4x + 2$, if x increases to $(x + \delta x)$, y increases to $(y + \delta y)$ and if the two points with coordinates (x, y) and $(x + \delta x, y + \delta y)$ lie on the line (Fig 5.4) we must have

$$y = 4x + 2 \tag{1}$$

$$y + \delta y = 4(x + \delta x) + 2 \tag{2}$$

115

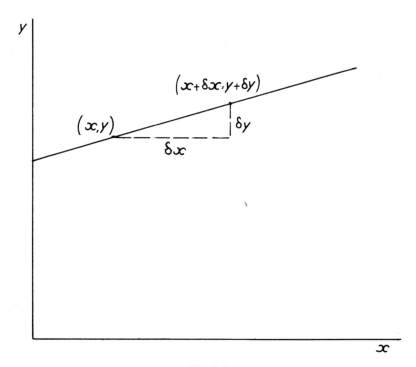

Fig 5.4

Subtracting eq. (1) from eq. (2), we obtain

$$(y + \delta y) - y = 4(x + \delta x) + 2 - (4x + 2)$$

$$\delta y = 4x + 4\ \delta x + 2 - 4x - 2$$

$$= 4\ \delta x$$

$$\therefore \quad \frac{\delta y}{\delta x} = 4$$

and

$$\frac{dy}{dx} = \lim_{\delta x \to 0} \left(\frac{\delta y}{\delta x} \right) = 4$$

This is exactly the expected result, because we know that the slope is equal to 4 for all positions along the line. That is, it is independent of the value of x.

In the case of a quadratic equation the same procedure can be adopted.

For example if

$$y = x^2 + 2x + 1 \qquad (3)$$

then $$y + \delta y = (x + \delta x)^2 + 2(x + \delta x) + 1 \qquad (4)$$

Subtracting eq. (3) from eq. (4)

$$y + \delta y - y = (x + \delta x)^2 + 2(x + \delta x) + 1 - (x^2 + 2x + 1)$$

$$= x^2 + 2x\,\delta x + (\delta x)^2 + 2x + 2\,\delta x + 1 - x^2 - 2x - 1$$

$$= 2x\,\delta x + (\delta x)^2 + 2\,\delta x$$

$$\therefore \quad \delta y = (2x + 2)\,\delta x + (\delta x)^2$$

$$\frac{\delta y}{\delta x} = (2x + 2) + \delta x$$

Now as δx decreases in size the second term on the right-hand side of this equation becomes of less and less significance, and in the limit

$$\frac{dy}{dx} = \lim_{\delta x \to 0} \left(\frac{\delta y}{\delta x} \right) = 2x + 2$$

This means that the slope of the curve at any point is a function of the point itself. For example,

when $x = 1$ $$\frac{dy}{dx} = 2 + 2 = 4$$

when $x = 2$ $$\frac{dy}{dx} = 4 + 2 = 6$$

5.2 SOME GENERAL RULES

The above procedure can be used to develop general rules for obtaining derivatives. For example in the case of the general quadratic

$$y = a + bx + cx^2$$
$$(y + \delta y) = a + b(x + \delta x) + c(x + \delta x)^2$$
$$(y + \delta y) - y = a + bx + b\ \delta x + cx^2 + 2cx\ \delta x + c(\delta x)^2 - (a + bx + cx^2)$$
$$\therefore \quad \delta y = b\ \delta x + 2cx\ \delta x + c(\delta x)^2$$
$$\frac{\delta y}{\delta x} = b + 2cx + c\ \delta x$$
$$\therefore \quad \frac{dy}{dx} = b + 2cx$$

that is, if $y = a + bx + cx^2$, then $dy/dx = b + 2cx$.

The derivative is in fact the sum of the derivatives of the three terms taken separately: that is,

if
$$y = a + bx + cx^2$$

then
$$\frac{dy}{dx} = \frac{d}{dx}(a) + \frac{d}{dx}(bx) + \frac{d}{dx}(cx^2)$$

and in general the derivative of any polynomial in x is equal to the sum of the derivatives of each term considered separately.

It can be seen that the derivative of the constant a is zero, that of the term bx is b and that of cx^2 becomes $2cx$. In general the derivative of x^n is equal to nx^{n-1}, which can be proved as follows:

If
$$y = ax^n,$$
$$y + \delta y = a(x + \delta x)^n$$
$$= a(x^n + {}^nC_1 x^{n-1}\ \delta x + \cdots + {}^nC_n(\delta x)^n)$$

from the binomial expansion (Section 4.11).
$$y + \delta y - y = a[x^n + {}^nC_1 x^{n-1}\ \delta x + \cdots + {}^nC_n(\delta x^n)] - ax^n$$

or
$$\delta y = a[{}^nC_1 x^{n-1}\ \delta x + {}^nC_2 x^{n-2}\ \delta x^2 + \cdots]$$

Hence
$$\frac{\delta y}{\delta x} = a^n C_1 x^{n-1} + {}^nC_2 x^{n-2}\ \delta x + \cdots$$

$$\therefore \quad \frac{dy}{dx} = \lim_{\delta x \to 0} \frac{\delta y}{\delta x} = anx^{n-1} + 0$$

So if $y = ax^n$, $dy/dx = anx^{n-1}$.

Example 1   If $y = 4x^6$, $n = 6$, $a = 4$, so

$$\frac{dy}{dx} = 4(6)x^5 = 24x^5$$

Example 2   If $y = x^{-2}$, $n = -2$, $a = 1$, so

$$\frac{dy}{dx} = (-2)x^{-3} = \frac{-2}{x^3}$$

THE PRODUCT RULE

If $y = uv$, where u and v are both functions of x, then the derivative

$$\frac{dy}{dx} = v\frac{du}{dx} + u\frac{dv}{dx}$$

This is obtained by the same method as used above. Since u and v are functions of x, a change in the value of x to $x + \delta x$ changes both u and v. Let the new values be $u + \delta u$ and $v + \delta v$. The value of y also changes, and we let this change be δy.

Hence, $\qquad\qquad\qquad y = uv$

and $\qquad\qquad\qquad y + \delta y = (u + \delta u)(v + \delta v)$

or $\qquad\qquad\qquad y + \delta y = uv + v\,\delta u + u\,\delta v + \delta u\,\delta v$

so that $\qquad\qquad\quad \delta y = v\,\delta u + u\,\delta v + \delta u\,\delta v$

and $\qquad\qquad\qquad \dfrac{\delta y}{\delta x} = v\dfrac{\delta u}{\delta x} + u\dfrac{\delta v}{\delta x} + \dfrac{\delta u}{\delta x}\,\delta v$

Now as $\delta x \to 0$, $\qquad \lim\dfrac{\delta y}{\delta x} = \dfrac{dy}{dx}$, $\quad \lim\dfrac{\delta u}{\delta x} = \dfrac{du}{dx}$

$$\lim\frac{\delta v}{\delta x} = \frac{dv}{dx}, \quad \lim \delta v \to 0$$

Taking limits of the expression gives

$$\frac{dy}{dx} = v\frac{du}{dx} + u\frac{dv}{dx} + \frac{du}{dx}(0) = v\frac{du}{dx} + u\frac{dv}{dx}$$

119

An example of the use of this is now given.

$$y = (x+2)(x^2+3)$$

Let $u = x+2$ and $v = x^2+3$

then $\qquad \dfrac{du}{dx} = 1 \qquad$ and $\qquad \dfrac{dv}{dx} = 2x$

so that $\qquad \dfrac{dy}{dx} = (x^2+3)(1) + (x+2)(2x)$

$$= x^2 + 3 + 2x^2 + 4x = 3x^2 + 4x + 3$$

THE QUOTIENT RULE

If $y = u/v$ where both u and v are functions of x, then

$$\frac{dy}{dx} = \frac{v(du/dx) - u(dv/dx)}{v^2}$$

This can be shown by allowing x to change to $x + \delta x$, and letting the corresponding changes in u, v and y be δu, δv and δy. Then

$$y = \frac{u}{v}$$

$$y + \delta y = \frac{u + \delta u}{v + \delta v}$$

Subtracting, we have

$$\delta y = \frac{u + \delta u}{v + \delta v} - \frac{u}{v} = \frac{vu + v\,\delta u - uv - u\,\delta v}{(v + \delta v)v}$$

$$= \frac{v\,\delta u - u\,\delta v}{v^2 + v\,\delta v}$$

Dividing by δx, we obtain

$$\frac{\delta y}{\delta x} = \frac{v(\delta u/\delta x) - u(\delta v/\delta x)}{v^2 + v\,\delta v}$$

As $\delta x \to 0$, $\qquad \dfrac{\delta y}{\delta x} \to \dfrac{dy}{dx}, \qquad \dfrac{\delta u}{\delta x} \to \dfrac{du}{dx}, \qquad \dfrac{\delta v}{\delta x} \to \dfrac{dv}{dx}, \qquad v\,\delta v \to 0.$

120

Hence $$\frac{dy}{dx} = \frac{v(du/dx) - u(dv/dx)}{v^2}$$

Example

$$y = \frac{x+2}{2x}$$

Let $u = x + 2$ and $v = 2x$

$$\frac{du}{dx} = 1 \qquad \frac{dv}{dx} = 2$$

$$\frac{dy}{dx} = \frac{(2x)(1) - (x+2)(2)}{(2x)^2}$$

$$= \frac{2x - 2x - 4}{4x^2} = \frac{-4}{4x^2} = \frac{-1}{x^2}$$

THE FUNCTION OF A FUNCTION OR CHAIN RULE

This rule states that if y is a function of u, and u is a function of x,

then $$\frac{dy}{dx} = \left(\frac{dy}{du}\right)\left(\frac{du}{dx}\right)$$

The proof of this rule is as follows

Let $$y = f(u) \qquad \text{and} \qquad u = g(x)$$

then $$y + \delta y = f(u + \delta u) \qquad \text{and} \qquad u + \delta u = g(x + \delta x)$$

Hence $$\frac{\delta y}{\delta x} = \frac{f(u + \delta u) - f(u)}{\delta x}$$

$$= \left(\frac{f(u + \delta u) - f(u)}{\delta u}\right)\frac{\delta u}{\delta x}$$

and $$\lim_{\delta x \to 0}\left(\frac{\delta y}{\delta x}\right) = \frac{dy}{dx} = \left(\frac{dy}{du}\right)\left(\frac{du}{dx}\right)$$

Example

$$y = (1 + x)^4$$

121

In this case, if $u = 1 + x$, then we have $y = u^4$ and $u = 1 + x$, so that y is a function of u and u is a function of x.

Now
$$\frac{dy}{du} = 4u^3 \qquad \frac{du}{dx} = 1$$

$$\frac{dy}{dx} = \frac{dy}{du}\frac{du}{dx} = (4u^3)(1) = 4(1 + x)^3$$

LOGARITHMS

Functions involving logarithms can be differentiated using some of the rules of Section 4.14. Initially we will not specify the base to which the logarithm is taken. To differentiate $y = \log x$, we allow x to increase by δx and let the corresponding increase in y be δy so that

$$y + \delta y = \log (x + \delta x)$$

Subtracting, we obtain

$$y + \delta y - y = \log (x + \delta x) - \log (x)$$

or
$$\delta y = \log \frac{(x + \delta x)}{x} = \log \left(1 + \frac{\delta x}{x} \right)$$

Now we divide by δx, and also multiply the term on the right-hand side by (x/x) or by 1. The latter leaves the value of the right-hand side unchanged.

$$\frac{\delta y}{\delta x} = \frac{1}{\delta x} \cdot \frac{x}{x} \log \left(1 + \frac{\delta x}{x} \right) = \left(\frac{x}{\delta x} \right) \frac{1}{x} \log \left(1 + \frac{\delta x}{x} \right)$$

$$= \frac{1}{x} \left[\log \left(1 + \frac{\delta x}{x} \right)^{x/\delta x} \right]$$

Let us consider the behaviour of the expression in the outer brackets. It is convenient to let $n = x/\delta x$. Since x is fixed, as $\delta x \to 0$, $n \to \infty$

Now
$$\frac{\delta y}{\delta x} = \frac{1}{x} \left[\log \left(1 + \frac{1}{n} \right)^n \right]$$

Taking the limit as $\delta x \to 0$ gives

$$\frac{dy}{dx} = \frac{1}{x} \log \left[\lim_{n \to \infty} \left(1 + \frac{1}{n} \right)^n \right]$$

But the limiting value of this expression is defined as e, and so

$$\frac{dy}{dx} = \frac{1}{x} \log (e)$$

If we now specify that we are using logarithms to the base e, since $\log (e) = 1$ we have

$$\frac{dy}{dx} = \frac{1}{x} \qquad \text{when} \qquad y = \log_e (x).$$

In general, from using the function of a function rule,

if
$$y = \log_e (u)$$

$$\frac{dy}{dx} = \frac{1}{u} \frac{du}{dx}$$

since
$$\frac{dy}{du} = \frac{1}{u}.$$

Example

$$y = \log_e (2x + 3)$$

$$u = 2x + 3, \qquad du/dx = 2$$

Hence
$$\frac{dy}{dx} = \frac{1}{(2x + 3)} (2) = \frac{2}{2x + 3}$$

We will assume for the rest of this book that whenever we are concerned with differentiation, logarithms use the base e. If this were not the case, a factor $\log_n (e)$ would need to be included, where n is the base of the logarithms.

EXPONENTIAL FUNCTIONS

These can be differentiated most easily by converting to logarithms.

For example, $y = e^x$

Taking logarithms, we obtain

$$\log y = \log (e^x) = x \log e = x$$

The derivative of $\log y$ is

$$\frac{d}{dx}(\log y) = \frac{1}{y}\left(\frac{dy}{dx}\right)$$

This arises from the function of a function rule, since y is a function of x, and $\log y$ is a function of y.

Hence, $$\frac{1}{y}\frac{dy}{dx} = 1$$

or $$\frac{dy}{dx} = y$$

This shows that if $y = e^x$, $dy/dx = e^x$ which is a rather surprising result. The same result is obtained by differentiating the series expansion of e^x (Section 4.12) term by term.

Other exponential functions can be differentiated by the same method. In particular,

if $$y = e^{ax}$$

$$\log y = ax \log e = ax$$

Differentiating, we have

$$\frac{1}{y}\frac{dy}{dx} = a$$

Hence $$\frac{dy}{dx} = ay = ae^{ax}$$

Example

$$y = e^{4x}$$

We have $a = 4$, and so

$$\frac{dy}{dx} = 4e^{4x}$$

TRIGONOMETRIC FUNCTIONS

A number of these are differentiated from first principles in Appendix A, Section 5 and we summarise the results in Table 5.1.

TABLE 5.1

Summary table of derivatives

y	$\dfrac{dy}{dx}$	y	$\dfrac{dy}{dx}$
a	0	$\log u$	$\dfrac{1}{u}\dfrac{du}{dx}$
ax^n	anx^{n-1}	e^{ax}	ae^{ax}
uv	$v\dfrac{du}{dx}+u\dfrac{dv}{dx}$	$\sin ax$	$a\cos ax$
$\dfrac{u}{v}$	$\dfrac{v\,(du/dx)-u(dv/dx)}{v^2}$	$\cos ax$	$-a\sin ax$
$f(u)$	$\dfrac{dy}{du}\dfrac{du}{dx}$	$\tan ax$	$\dfrac{a}{\cos^2 ax}$

a is a constant, u and v are functions of x.

5.3   EXERCISES

Find dy/dx for the following functions:

1. $y = 3x - 4$

2. $y = x^2 - 3x + 3$

3. $y = 3x^4 - x^3 + x^2 + 25x$

4. $y = 6x^3 - \dfrac{x^4}{4} + \dfrac{x^3}{3} - \dfrac{1}{x^2}$

5. $y = (2x^2 + 3x - 1)\left(\dfrac{1}{x} - 3x + 2\right)$

6. $y = (2x + 4)\left(3x + \dfrac{2}{x^2}\right)$

125

7. $y = \dfrac{4x^2 + 4}{(x^2 + 3x + 2)}$

8. $y = (2x + 3)^5$

9. $y = \log (2x + 3)$

10. $y = x \log (2x^2 + 3x - 5)$

11. $y = 3e^{2x} + 4e^{-3x} + x^2 e^x$

12. $y = \sin 2x + 3 \cos 5x - \tan 3x$.

5.4 ELASTICITY OF DEMAND

The quantity demanded of some goods is much more sensitive to changes in price than is the quantity demanded of others and this difference is clearly very important. A measure of sensitivity to price change can be obtained by expressing the percentage change in the quantity demanded of a good in terms of the percentage change in price. For very small changes in price this ratio is the *price elasticity of demand*:

$$E_D = \frac{(dq/q) \times 100}{(dp/p) \times 100} = \frac{dq}{dp} \frac{p}{q}$$

If the demand curve is known in the form $q = f(p)$, then it is a simple matter to differentiate and to substitute the values of dq/dp in the above equation. For example, if

$$q = f(p) = 100 - p - p^2$$

$$\frac{dq}{dp} = -1 - 2p = -(1 + 2p)$$

$$\therefore \quad E_D = -(1 + 2p)\frac{p}{q}$$

If the present price is £5 then the quantity demanded is

$$q = 100 - 5 - 25 = 70$$

and

$$E_D = -(1 + 10)\frac{5}{70}$$

$$= -0.8 \text{ approximately}$$

This means that the demand decreases by approximately 0.8 per cent for a 1 per cent increase in price or, conversely, increases by 0.8 per cent for a 1 per cent decrease in price. It follows that, in this case, a decrease in price results in a decrease in total revenue because

$$\text{total revenue} = \text{price} \times \text{quantity} = pq$$

In this case the demand is said to be *inelastic* at a price of £5.

If, however, the present price is £6,

then
$$q = 100 - 6 - 36 = 58$$

and
$$E_D = -(1 + 12)\frac{6}{58}$$

$$= -1.3 \text{ approximately}$$

Thus a 1 per cent decrease in price results in a 1.3 per cent increase in the quantity demanded and if this is supplied it results in an increase in the total revenue. The demand is then said to be *elastic* at a price of £6.

When the elasticity of demand is unity, a rise or fall in price results in the same percentage decrease or increase in quantity demanded and, therefore, leaves the total revenue the same.

Other coefficients of elasticities are of importance in economics and these are determined in a similar way to the above, eg we define

$$\textit{income elasticity of demand} = \frac{\text{percentage change in quantity demanded}}{\text{percentage change in income}}$$

which can be expressed mathematically as

$$\frac{(dq/q) \times 100}{(dI/I) \times 100} \quad \text{or} \quad \frac{dq}{dI}\frac{I}{q}$$

To determine this elasticity it is necessary to know the relationship between the quantity demanded of a good and the income, that is,

$$q = f(I)$$

It is, therefore, possible to determine other forms of elasticity if the relationship between the variables is known.

127

5.5 MARGINAL ANALYSIS

We have, in the differential calculus, a method for determining the slope of a number of functions. In terms of the total cost curve example, the slope at any point tells us the marginal or extra cost incurred if the volume of production is increased by a very small amount or, conversely, the cost saved if the volume of production is decreased by a very small amount.

The same principles also apply in other uses of marginal concepts, such as marginal revenue (obtained from a total revenue function), the marginal productivity of labour (obtained from a production function), and the marginal propensity to consume (obtained from a consumption function).

For the linear cost function

$$y = 100 + 3x \qquad \frac{dy}{dx} = 3$$

The marginal cost is equal to 3 and is independent of the volume of production.

This can be checked by calculating the cost at any two levels of production which are one unit apart. For example,

when	$x = 3,$	$y = 100 + 9 = 109$
	$x = 4,$	$y = 100 + 12 = 112$
or when	$x = 15,$	$y = 100 + 45 = 145$
	$x = 16,$	$y = 100 + 48 = 148$

Both these sets of costs differ by 3 units.

For the quadratic function

$$y = 100 + 2x + \tfrac{1}{10}x^2$$

$$\frac{dy}{dx} = 2 + \tfrac{1}{5}x$$

In this case the marginal cost is a function of the output. For example,

when $\qquad x = 50, \qquad \dfrac{dy}{dx} = 2 + 10 = 12$

and when $\qquad x = 100, \qquad \dfrac{dy}{dx} = 2 + 20 = 22$

Therefore, the marginal cost is positive and increasing as output increases.

However, if, as is often the case, more efficient use can be made of the variable factors of production as the level of output is increased, then it is reasonable to expect the marginal cost to be positive but to decrease with output. To satisfy this requirement the derivative could be of the form

$$\frac{dy}{dx} = a - bx$$

where a and b are positive constants. Then as x is increased the marginal cost decreases.

Before going on to consider maximum and minimum values we end this section on marginal analysis by relating marginal revenue and elasticity of demand.

If $TR = pq$, where TR is total revenue, and q is a function of p, then differentiating gives

$$\frac{dTR}{dq} = p + q\frac{dp}{dq} = p\left(1 + \frac{q}{p}\frac{dp}{dq}\right)$$

But marginal revenue is given by

$$MR = \frac{dTR}{dq}$$

and elasticity of demand

$$E_D = \frac{p}{q}\frac{dq}{dp}$$

and hence

$$MR = p\left(1 + \frac{1}{E_D}\right)$$

When $E_D = -1$, $MR = 0$, that is, revenue is constant for changes in q. When $E_D = \infty$, $MR = p$, that is, when the firm is a competitor in the product market and the demand is perfectly elastic, the marginal revenue equals price.

5.6 EXERCISES

1. Find the marginal-cost and average-cost function from the following total-cost functions:

   (a) $TC = 4q^3 + 2q^2 - 25q$

   (b) $TC = (q^3 - 3q)(16 + 5q)$

   (c) $TC = 25 + 6qe^{2q}$

   (d) $TC = (3 \log q + 5)(q^2 - q)$

2. Find the elasticity of demand when $p = 10$ for the following equations:

Demand

$p + 2q = 50$

$qp^2 = 400$

$q - 35 = -3p$

3. Show that the elasticity of demand at any point on the curve $q = cp^{-1}$ is -1, on the curve $q = cp^{-4}$ is -4, and in general that the elasticity of demand at any point on the curve $q = cp^{-n}$ is $-n$. (Such curves are known as *constant-elasticity* demand curves.)

5.7 MAXIMA AND MINIMA

A total-cost function of the form

$$y = 100 + 2x + \tfrac{1}{10}x^2$$

tells us the total cost of manufacture in terms of the total output. The average cost per unit can be obtained by dividing throughout by the quantity produced:

$$\text{average cost per unit} = \frac{\text{total cost}}{\text{quantity}} = \frac{100 + 2x + \tfrac{1}{10}x^2}{x}$$

$$= \frac{100}{x} + 2 + \tfrac{1}{10}x$$

This is an equation showing that the average cost is very high when only a small number of units are made. Initially this value decreases with an increase in output, but eventually the increasing marginal costs cause the average cost per unit to rise. There is, therefore, some value of output for which the average cost per unit is a minimum.

A rough sketch of the graphs of the functions shows up the situation more clearly (Fig 5.5).

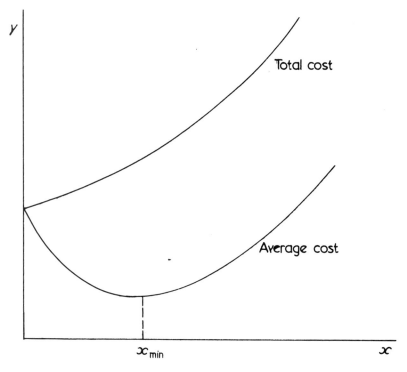

Fig 5.5

Then for

$$x < x_{min} \qquad \frac{dAC}{dx} < 0 \qquad \text{(ie the slope is negative)}$$

$$x > x_{min} \qquad \frac{dAC}{dx} > 0 \qquad \text{(ie the slope is positive)}$$

$$x = x_{min} \qquad \frac{dAC}{dx} = 0$$

Therefore, the value of output for which the average cost per unit is a minimum can be found by the differential calculus.

$$AC = \frac{100}{x} + 2 + \tfrac{1}{10}x$$

$$= 100x^{-1} + 2 + \tfrac{1}{10}x$$

$$\therefore \quad \frac{d(AC)}{dx} = -100x^{-2} + \tfrac{1}{10} = \frac{-100}{x^2} + \tfrac{1}{10}$$

$$\frac{d(AC)}{dx} = 0 \qquad \text{when} \qquad -\frac{100}{x^2} + \tfrac{1}{10} = 0$$

that is, $\qquad \dfrac{100}{x^2} = \tfrac{1}{10}$

and $\qquad x^2 = 1,000 \qquad \text{or} \qquad x = \sqrt{(1,000)} = 31.623$

This is the value of output for which the average cost per unit is a minimum. For an output level of 31 or 32 units,

$$x = 31 \qquad \text{and} \qquad AC = \frac{100}{31} + 2 + \frac{31}{10} = 8.3258$$

$$x = 32 \qquad \text{and} \qquad AC = \frac{100}{32} + 2 + \frac{32}{10} = 8.3250$$

If it were possible to obtain an output of 31.623 units then the average cost per unit would be

$$AC = \frac{100}{31.623} + 2 + \frac{31.623}{10}$$

$$= 3.1623 + 2 + 3.1623 = 8.3246$$

The shape of the curve is, therefore, fairly flat about the minimum point and for most practical purposes the difference between the average cost per unit at these levels of output would be considered to be insignificant.

It is interesting to note that at the minimum the term $100/x$ is equal in value to the term $x/10$. The former is equal to the share of the fixed costs which must be carried by each unit of output and, therefore, decreases with output. The latter corresponds to that part of the marginal cost function which is dependent upon the quantity produced. One term, therefore, decreases with output and the other increases, and the minimum average cost occurs when the two terms are equal. At this point the average cost per unit is equal to the marginal cost.

$$\text{Total cost, } y = 100 + 2x + \tfrac{1}{10}x^2$$

$$\text{marginal cost, } \frac{dy}{dx} = 2 + \tfrac{1}{5}x$$

For an output of 31.623 units

$$\text{marginal cost} = 2 + \tfrac{1}{5} \times 31.623 = 8.3246$$

which is the value of the minimum average cost per unit.

This follows from the fact that for values of output less than x_{min} the average cost is falling and this can only happen if the cost of producing each extra unit is less than the average cost of producing all previous units; that is, when

$$0 < x < x_{min}, \qquad MC < AC$$

When the average cost is increasing, this is due to the fact that the cost of producing each extra unit is greater than the average cost of producing all previous units; that is, when

$$x > x_{min}, \qquad MC > AC$$

Then at the minimum the two must be equal; that is, when

$$x = x_{min}, \qquad MC = AC$$

In this example the minimum value was determined by setting the derivative equal to zero. For a curve which reaches a maximum value the same sort of arguments apply with the result that:

for $\qquad\qquad x < x_{max}$ the slope is positive

$\qquad\qquad\qquad x > x_{max}$ the slope is negative

$\qquad\qquad\qquad x = x_{max}$ the slope is zero

133

and at the maximum value the derivative is zero. It is therefore necessary to decide whether a point for which the derivative is zero corresponds to a maximum or a minimum value. For the average cost function the graph shows that there is a minimum value. However, it is not always easy to sketch the graph and it is not necessary since the differential calculus can be used to distinguish between values corresponding to maximum and minimum positions.

To do this it is simplest to differentiate the function a second time following the same procedure as was used to obtain the first derivative dy/dx. This can best be illustrated by reference to two examples.

In the case of the linear function

$$y = 100 + 3x, \qquad \frac{dy}{dx} = 3$$

This is constant for all values of x and a graph of dy/dx against x is as shown in Fig 5.6.

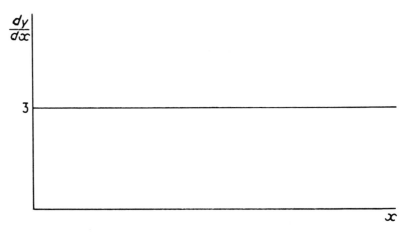

Fig 5.6

The slope of this line is zero at all points. The derivative of dy/dx is therefore zero. If we write y' for dy/dx, then

$$y' = \frac{dy}{dx} = 3 \qquad \text{and} \qquad \frac{dy'}{dx} = 0$$

The derivative of y' is then the second derivative of y and can be written in the alternative forms

$$\frac{dy'}{dx} = \frac{d}{dx}\left(\frac{dy}{dx}\right) = \frac{d^2y}{dx^2}$$

If we let $y'' = d^2y/dx^2$, then the third derivative can be formed in a similar way:

$$\frac{dy''}{dx} = y''' = \frac{d}{dx}\left(\frac{d^2y}{dx^2}\right) = \frac{d^3y}{dx^3}$$

and in general the nth derivative is

$$\frac{d^ny}{dx^n}$$

In the case of the quadratic equation

$$y = a + bx + cx^2$$

then

$$\frac{dy}{dx} = b + 2cx$$

$$\frac{d^2y}{dx^2} = 2c$$

and d^3y/dx^3 and all higher order derivatives are equal to zero.

In general a polynomial of degree n, ie an equation in which the highest power of the variable is n, has n derivatives not equal to zero and the $(n+1)$th is always equal to zero.

For the purposes of the analysis we can restrict ourselves to the first two derivatives because they are usually sufficient to determine whether a function has a maximum or minimum value and to distinguish between the two.

For the quadratic

$$y = a + bx + cx^2$$

the slope at any point of the curve is given by

$$\frac{dy}{dx} = b + 2cx$$

135

This will be positive or negative depending upon the values of b, c and x. Let us consider the two cases of the general quadratic equation over the whole range of possible values of x:

(a) when c is positive

(b) when c is negative

The general shape of these functions is shown in Fig 5.7.

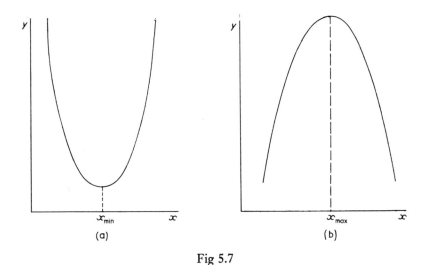

Fig 5.7

Fig 5.7(a) has a minimum point, ie for $x = x_{min}$ the function has a value which is lower than the value for all other values of x.

Fig 5.7(b) has a maximum point, ie for $x = x_{max}$ the function has a value which is higher than the value for all other values of x.

The slope of these curves depends upon the value of x and the sign of the slope depends on the sign of the constant c. This can be seen more clearly if the graph of dy/dx against x is drawn (Fig 5.8).

The slope of the curve at a minimum and a maximum point is zero and therefore the graph of dy/dx crosses the axis of x at this point. But in Fig 5.7 (a) for

$$x < x_{\min}, \qquad \frac{dy}{dx} < 0$$

$$x > x_{\min}, \qquad \frac{dy}{dx} > 0$$

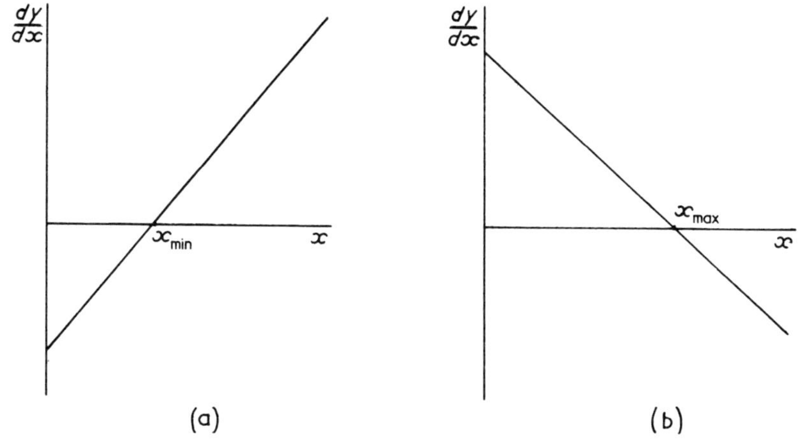

(a)   (b)

Fig 5.8

The slope of the graph of dy/dx against x is therefore positive. In Fig 5.7 (b), for

$$x < x_{\max}, \qquad \frac{dy}{dx} > 0$$

$$x > x_{\max}, \qquad \frac{dy}{dx} < 0$$

Therefore, the slope of dy/dx against x is negative.

The slope of these two graphs can, of course, be determined by differentiating the linear function

$$\frac{dy}{dx} = b + 2cx$$

137

with the result

$$\frac{d^2y}{dx^2} = 2c$$

The graph of this is illustrated in Fig 5.9 for both signs of the constant c.

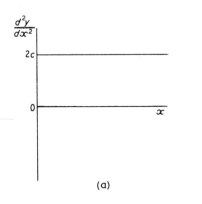
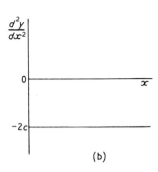

(a)

(b)

Fig 5.9

It follows that a quadratic function has a maximum or minimum at the point where

$$\frac{dy}{dx} = 0$$

and that this point is a maximum if d^2y/dx^2 is negative and a minimum if d^2y/dx^2 is positive.

The use of these rules is demonstrated in the following examples:

Example 1   Total cost $= y = 100 + 2x - \frac{1}{10}x^2$

$$\frac{dy}{dx} = 2 - \tfrac{1}{5}x$$

and

$$\frac{d^2y}{dx^2} = -\tfrac{1}{5}$$

138

This is negative and the function has a maximum at the point where

$$\frac{dy}{dx} = 2 - \tfrac{1}{5}x = 0$$

that is, where $x = 10$ and the maximum value of total cost is given by $y = 100 + 20 - \frac{100}{10} = 110$.

Example 2 Given that total cost $y = 100 + 2x + \tfrac{1}{10}x^2$, find the value of x which minimises the average cost, and hence find the minimum value.

$$\text{Average cost} = AC = \frac{100}{x} + 2 + \frac{x}{10}$$

$$\frac{d(AC)}{dx} = -\frac{100}{x^2} + \tfrac{1}{10} = -100x^{-2} + \tfrac{1}{10}$$

$$\frac{d^2(AC)}{dx^2} = (-100)(-2)x^{-3} = \frac{200}{x^3}$$

There is a maximum or a minimum when $d(AC)/dx = 0$, that is, when

$$\frac{-100}{x^2} + \tfrac{1}{10} = 0$$

or
$$x = 31.623$$

At this value of output

$$\frac{d^2(AC)}{dx^2} = \frac{200}{x^3} = \frac{200}{(31.623)^3}$$

This is positive and therefore the value of $x = 31.623$ gives the minimum value of the average cost function

$$AC = \frac{100}{31.623} + 2 + \frac{31.623}{10} = 8.3246$$

Example 3 The cubic equation

$$y = a + bx + cx^2 + dx^3$$

139

can have a number of forms depending upon the sign and size of the coefficients of the variable x. The graphs in Fig 5.10 are examples of two of the forms which it might take.

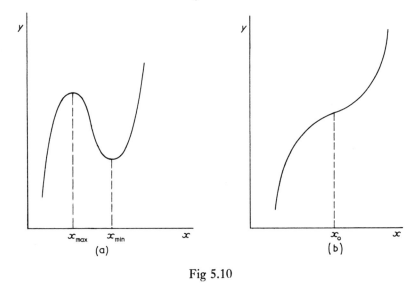

Fig 5.10

Fig 5.10 (a) has no true maximum or minimum value in terms of our previous definition because y increases indefinitely as x increases and y decreases indefinitely as x decreases. The values of the function corresponding to the value x_{max} and x_{min} are, however, called *local maximum and local minimum*. If we apply the graphical reasoning to this function that was applied to the quadratic function it is easily seen that these two points fulfil the conditions required for maxima and minima.

Fig 5.10 (b) has no local maximum or minimum values but has a value x_0 for which the function has a slope of zero. The point x_0, therefore, satisfies the condition $dy/dx = 0$. But it gives the value $d^2y/dx^2 = 0$, which does not satisfy the second-order conditions for a maximum or minimum. It is, in fact, called a *point of inflexion*, which means that $d^2y/dx^2 = 0$ there but dy/dx does not change sign as x increases through x_0.

From these results it is possible to define a set of rules.

(1) A function has a stationary value, which can be either a maximum, a minimum or a point of inflexion at the point where

$$\frac{dy}{dx} = 0$$

(2) (a) if d^2y/dx^2 is negative at that point, the stationary value is a maximum

(b) if d^2y/dx^2 is positive at that point the stationary value is a minimum

(c) if d^2y/dx^2 is zero the stationary value may be a maximum, minimum or point of inflexion and the curve should be sketched to examine it.

The condition (1) is frequently referred to as the *necessary* condition for a stationary value. That is, unless this is satisfied neither a maximum, nor minimum nor point of inflexion can occur. This condition is not, however, *sufficient* for a maximum, since it is also satisfied at a minimum or point of inflexion. Conditions (1) and (2) together make up the *necessary and sufficient* conditions for a maximum, minimum or point of inflexion since they include all the possibilities. If a stationary value occurs, it satisfies these conditions. If a value of x occurs which satisfies these conditions it is a stationary value.

5.8 COST AND REVENUE ANALYSIS

The differential calculus enables us to determine the stationary values of a function and to decide whether these values are maxima or minima. This important result is of frequent use in economics as the following example shows:

Example Let the total cost function be

$$c = 100 + 2q + \tfrac{1}{10}q^2$$

and let the demand function be

$$p = 20 - \tfrac{1}{8}q$$

141

Then the total revenue $TR = pq$. Substituting the above expression for p in the revenue function gives

$$TR = pq = (20 - \tfrac{1}{5}q)q$$
$$= 20q - \tfrac{1}{5}q^2$$

This is a quadratic equation and it can be sketched roughly by determining a few points on the curve.

1. $TR = 0$ when $20q - \tfrac{1}{5}q^2 = 0$

   that is, when $q(20 - \tfrac{1}{5}q) = 0$

   $$q = 0 \text{ or } 100$$

2. $\dfrac{d(TR)}{dq} = 20 - \tfrac{2}{5}q$

   This equals zero when

   $$\tfrac{2}{5}q = 20, \qquad q = 50$$

   and at this point

   $$TR = 20 \times 50 - \tfrac{1}{5} \times 50 \times 50 = 500$$

To check whether this is a maximum or a minimum it is necessary to obtain the second-order derivative

$$\frac{d^2(TR)}{dq^2} = -\tfrac{2}{5}$$

This is negative, and therefore the maximum total revenue occurs at an output of 50 units.

The curve is approximately as in Fig 5.11. Superimposing the cost function on the same graph indicates that the net revenue NR is given by the vertical distance between the two curves:

net revenue = total revenue − total cost of production

$$NR = TR - C$$

This is positive over the range $q = q_1$ to $q = q_2$. It is not, however, necessarily at its maximum at the output where total revenue is at its

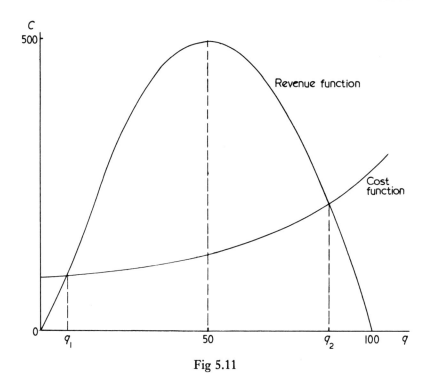

Fig 5.11

maximum, because costs are increasing with output. To determine the output where net revenue is a maximum use can be made of the differential calculus:

$$NR = TR - C$$
$$= (20q - \tfrac{1}{5}q^2) - (100 + 2q + \tfrac{1}{10}q^2)$$
$$= 18q - \tfrac{3}{10}q^2 - 100$$
$$\therefore \quad \frac{d(NR)}{dq} = 18 - \tfrac{3}{5}q$$

This is zero when

$$\tfrac{3}{5}q = 18 \quad \text{or} \quad q = 30$$

143

At this output

$$NR = 18 \times 30 - \tfrac{3}{10} \times 30 \times 30 - 100 = 170$$

To check whether this is a maximum or a minimum we differentiate a second time with the result

$$\frac{d^2(NR)}{dq^2} = -\tfrac{3}{5}$$

This is negative and, therefore, this net revenue is a maximum. The maximum gross revenue occurs when $q = 50$. At this output

$$NR = 18 \times 50 - \tfrac{3}{10} \times 50 \times 50 - 100 = 50$$

This is considerably less than at the reduced output of 30. Finally,

$$\frac{d(TR)}{dq} = 20 - \tfrac{2}{5}q \qquad \text{and} \qquad \frac{dC}{dq} = 2 + \tfrac{1}{5}q$$

Therefore at an output of 30

$$\text{Marginal revenue} = \frac{d(TR)}{dq} = 20 - \tfrac{2}{5} \cdot 30 = 8$$

$$\text{Marginal cost} = \frac{dC}{dq} = 2 + \tfrac{1}{5} \cdot 30 = 8$$

This confirms that maximum net revenue is obtained when

$$\text{marginal revenue} = \text{marginal cost}$$

and
$$\text{for } q < 30, \qquad MR > MC$$
$$\text{for } q > 30, \qquad MR < MC$$

5.9 EXERCISES

1. Determine the stationary values of the following functions:
   (a) $y = 3x^2 - 120x + 30$
   (b) $y = 16 - 8x - x^2$
   (c) $y = x^4$
   (d) $y = (x - 5)^3$

2. The total cost of producing an output q is given by

$$c = 500 + 4q + \tfrac{1}{2}q^2$$

Obtain the value of q which minimises the average cost.

3. If the demand function for a monopolist's product is $q + 2p = 10$, determine the price and output required to maximise the total revenue. What is the elasticity of demand at this price?

4. The total cost function for a product is $TC = q^3 - 20q^2 + 20$ and the demand function is $q + 2p = 24$. Find the price and level of output required to maximise a monopolist's net revenue.

5.10 DIFFERENTIALS

The derivative dy/dx represents the rate of change of y with respect to x for very small changes in x. It is defined as the *ratio*

$$\frac{dy}{dx} = \lim_{\delta x \to 0} \left(\frac{\delta y}{\delta x} \right)$$

where δy and δx are small but finite changes in y and x. The relationship between these is shown in Fig 5.12.

In fact,
$$\frac{\delta y}{\delta x} = \frac{dy}{dx} + \epsilon$$

or
$$\delta y = \left(\frac{dy}{dx} \right) \delta x + \epsilon \, \delta x$$

where $\epsilon \to 0$ as Q moves close to P. When this occurs, $\delta y \to dy$ $\delta x \to dx$ and $\epsilon \, \delta x \to 0$ so that

$$dy = \left(\frac{dy}{dx} \right) dx$$

We define dy as the *differential* of y and dx the differential of x. It follows that

$$\frac{dy}{dx} = \frac{\text{differential of } y}{\text{differential of } x} = \lim_{\delta x \to 0} \left(\frac{\delta y}{\delta x} \right)$$

Using this notation it is possible to calculate the change which will occur in y for a small change in x when y is a function of x. For

145

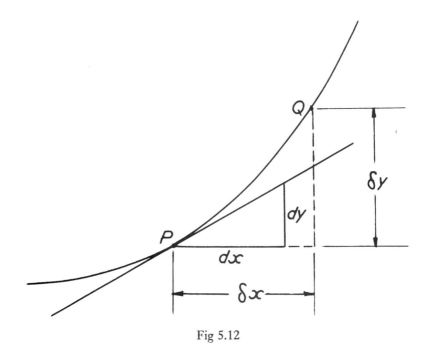

Fig 5.12

example if $y = 3x^2 + 2x + 1$, then $dy/dx = 6x + 2$ and $dy = (6x + 2)\, dx$ which means that the change in y which occurs due to a small change in x depends upon the value of x at which the change takes place.

If $x = 10$ then $dy = [(6 \times 10) + 2)]\, dx = 62\, dx$

If $x = 1$ then $dy = (6 + 2)\, dx = 8\, dx$

Therefore, the change in y resulting from a small change in x becomes larger as x increases in value.

5.11 TAYLOR'S AND MACLAURIN'S THEOREMS

These are two theorems which have many important applications in advanced economic theory. Here they will be used to obtain some important series expansions. Taylor's theorem states that, for a general function f,

146

$$f(a+x)=f(a)+xf'(a)+\frac{x^2}{2!}f''(a)+\cdots+\frac{x^n}{n!}f^{(n)}(a)+\cdots$$

if the series converges, where

$f(a)$ is the value of the function $f(a+x)$ when $x=0$,

$f'(a)$ is the value of $f'(a+x)$ when $x=0$

$f^{(n)}(a)$ is the value of the nth derivative of $f(a+x)$ when $x=0$

As an illustration of the use of Taylor's theorem let us consider the function $\log(1+x)$

Let $\qquad\qquad f(a+x)=\log(1+x)$

Then $\qquad\qquad f(a)=\log(1+0)=0$

$$f'(a+x)=\frac{1}{1+x} \qquad \text{and} \qquad f'(a)=1$$

$$f''(a+x)=\frac{-1}{(1+x)^2} \qquad \text{and} \qquad f''(a)=-1$$

$$f'''(a+x)=\frac{2!}{(1+x)^3} \qquad \text{and} \qquad f'''(a)=2!$$

$$f^{(4)}(a+x)=\frac{-3!}{(1+x)^4} \qquad \text{and} \qquad f^{(4)}(a)=-3!$$

so that

$$\log(1+x)=0+x-\frac{x^2}{2!}+\frac{x^3}{3!}(2!)+\frac{x^4}{4!}(-3!)+\cdots$$

$$=x-\frac{x^2}{2}+\frac{x^3}{3}-\frac{x^4}{4}+\frac{x^5}{5}+\cdots$$

which is an infinite series. It can be shown that this series converges provided $-1<x<+1$.

MacLaurin's theorem can be regarded as the special case of Taylor's theorem which occurs when $a=0$. It states

$$f(x)=f(0)+xf'(0)+\frac{x^2}{2!}f''(0)+\cdots+\frac{x^n}{n!}f^{(n)}(0)+\cdots$$

147

if the series converges. The derivatives are evaluated at $x = 0$.

Example 1

$$f(x) = \sin x \qquad f(0) = \sin 0 = 0$$

$$f'(x) = \cos x \qquad f'(0) = \cos 0 = 1$$

$$f''(x) = -\sin x \qquad f''(0) = -\sin 0 = 0$$

$$f'''(x) = -\cos x \qquad f'''(0) = -\cos 0 = -1$$

This process can be repeated indefinitely and an infinite series of terms is obtained of which all the odd terms are zero and the even terms are alternately $+1$ and -1. Applying MacLaurin's theorem, we obtain

$$f(x) = \sin x = 0 + x(1) + \frac{x^2}{2!}(0) + \frac{x^3}{3!}(-1) + \cdots$$

$$= x - \frac{x^3}{3!} + \frac{x^5}{5!} - \frac{x^7}{7!} + \cdots$$

Sin x is known as an odd function of x because its expansion contains only odd powers of x.

MacLaurin's theorem can also be used to expand the functions $\cos x$ and e^x

Example 2

$$f(x) = \cos x \qquad f(0) = \cos 0 = 1$$

$$f'(x) = -\sin x \qquad f'(0) = -\sin 0 = 0$$

$$f''(x) = -\cos x \qquad f''(0) = -\cos 0 = -1$$

$$f'''(x) = \sin x \qquad f'''(0) = \sin 0 = 0$$

Extending this process further and substituting the results in MacLaurin's expansion gives

$$\cos x = 1 - \frac{x^2}{2!} + \frac{x^4}{4!} - \frac{x^6}{6!} + \cdots$$

Cos x is known as an even function of x because its expansion contains only even powers of x.

Example 3

$$f(x) = e^x \qquad f(0) = e^0 = 1$$
$$f'(x) = e^x \qquad f'(x) = e^0 = 1$$
$$f''(x) = e^x \qquad f''(0) = e^0 = 1$$
$$f'''(x) = e^x \qquad f'''(0) = e^0 = 1$$

which can be continued to give

$$e^x = 1 + x + \frac{x^2}{2!} + \frac{x^3}{3!} + \cdots$$

5.12 EULER RELATIONS

These arise from the three series which were obtained using MacLaurin's theorem. Since

$$e^x = 1 + x + \frac{x^2}{2!} + \frac{x^3}{3!} + \frac{x^4}{4!} + \cdots$$

if the index is multiplied by the imaginary number i (see Section 3.5) and the function expanded as an infinite series, we obtain

$$e^{ix} = 1 + (ix) + \frac{(ix)^2}{2!} + \frac{(ix)^3}{3!} + \frac{(ix)^4}{4!} + \cdots$$

and remembering that $i = \sqrt{(-1)}$, $i^2 = -1$, $i^3 = -i$, $i^4 = 1$, etc, we have

$$e^{ix} = 1 + ix - \frac{x^2}{2!} - \frac{ix^3}{3!} + \frac{x^4}{4!} + \cdots$$

This can be separated into real and imaginary parts as follows:

$$e^{ix} = \left(1 - \frac{x^2}{2!} + \frac{x^4}{4!} - \cdots\right) + i\left(x - \frac{x^3}{3!} + \frac{x^5}{5!} - \cdots\right)$$

$$= \cos x + i \sin x.$$

By similar reasoning it can be shown that

$$e^{-ix} = \cos x - i \sin x.$$

These relationships are important and will be found extremely useful in the treatment of certain differential equations (see Section 9.4).

149

5.13 NEWTON'S METHOD

It is possible to use the differential calculus to determine the real roots of any polynomial in x. This is particularly useful in the case of cubic and higher-order equations when the roots are not integer values.

For example let the cubic equation be

$$x^3 + 2x^2 + x - 5 = 0$$

We can determine a very approximate value for a root to this equation by substituting various integer values for the variable. For example,

$$\text{let } x = 0 \qquad \text{then} \qquad f(x) = 0 + 0 + 0 - 5 = -5$$
$$x = 1 \qquad\qquad f(x) = 1 + 2 + 1 - 5 = -1$$
$$x = 2 \qquad\qquad f(x) = 8 + 8 + 2 - 5 = 13$$

With this information we can see that the function changes sign between the values $x = 1$ and $x = 2$. It follows that at least one root of the cubic equation lies between these two values. One of these values is taken as a starting point from which it is possible to proceed in a series of steps to the actual solution. The procedure can be stated, in general, as follows.

Let a be one of the two consecutive integer values between which the function changes sign. Then the true root of the equation is $a + h$, where h is an unknown whose absolute value lies between 0 and 1. If a is the lower of the two integers then h is positive whereas if a is the higher integer h is negative.

Taylor's theorem states

$$f(a+h) = f(a) + hf'(a) + \frac{h^2}{2!} f''(a) + \cdots$$

If h is small then approximately

$$f(a+h) = f(a) + hf'(a)$$

Now for this to be a root, $f(a+h) = 0$ and hence

$$h_1 = \frac{-f(a)}{f'(a)} \tag{1}$$

150

is close to h. Thus a first approximation to the root $a+h$ is given by $a+h_1$ where a is a known value and h_1 is $-f(a)/f'(a)$.

In the example we are considering

$$f(x) = x^3 + 2x^2 + x - 5 \quad \text{and} \quad a = 1$$

$$\therefore \quad f(a) = f(1) = 1 + 2 + 1 - 5 = -1$$

Also
$$f'(x) = \frac{d}{dx}\{f(x)\} = 3x^2 + 4x + 1$$

$$\therefore \quad f'(a) = f(1) = 3 + 4 + 1 = 8.$$

Substituting these values in eq. (1)

$$h_1 = \frac{-f(1)}{f'(1)} = \frac{-(-1)}{8} = \frac{1}{8} = 0.125$$

At this stage in the procedure a better approximation to the root of the cubic equation is given by $(a+h_1)$, that is $(1+0.125) = 1.125$. Let us call this value a_1. Then we can approach even closer to the true root by letting this be denoted by (a_1+h_2) where

$$f(a_1 + h_2) = f(a_1) + h_2 f'(a_1) = 0$$

or
$$h_2 = \frac{-f(a_1)}{f'(a_1)}$$

As before $f(x) = x^3 + 2x^2 + x - 5$

$$\therefore \quad f(a_1) = f(1.125) = (1.125)^3 + 2(1.125)^2 + (1.125) - 5 = 0.0801$$

and $f'(a_1) = f'(1.125) = 3(1.125)^2 + 4(1.125) + 1 = 9.2969$

$$\therefore \quad h_2 = \frac{-f(1.125)}{f'(1.125)} = -\frac{0.0801}{9.2969} = -0.0086$$

A closer approximation to the root of the cubic equation is given by $(a_1 + h_2) = (1.125 - 0.0086) = 1.1164$, for which $f(1.1164) = 0.0005$.

This procedure can be repeated an indefinite number of times and successive estimates of the root of the equation will oscillate about the true value with ever decreasing amplitude. It can be carried out on a computer because of its repetitive stepwise approach. The roots of any polynomial in x for which the first and second derivatives exist (ie for any continuous function of x) can be obtained by this method.

5.14 EXERCISES

1. Use Taylor's theorem to expand $(1+x)^4$ and hence show that $1.5^4 = 5.0625$.
2. Use MacLaurin's theorem to expand $(1+x)^{-1}$ as a power series.
3. Use Newton's method to obtain a value of x which satisfies the following equations using the given approximate root as a starting point.

   (a) $x^2 + 4.55x - 8.70 = 0$

   Approximate root $x = 1.5$

   (b) $x^3 - 2.34x^2 + 2x - 4.68 = 0$

   Approximate root $x = 2.25$

   (c) $2x^3 - 8.2x^2 - 3x + 12.3 = 0$

   Approximate root $x = 4$

4. What rate of interest makes the present values of the following two projects equal?

Return at end of year	1	2	3	4
Project A	70	80	100	100
Project B	60	90	80	130

   (*Hint*: If i is the rate of interest and $R = 1 + i$ then the cubic equation in R can be solved using $R = 1$ as an approximate root.)

CHAPTER SIX

Integral Calculus

6.1 INTRODUCTION

If x is the level of output and TC is the total cost, then given a cost curve

$$TC = 3x^2 + 20x + 13$$

by differentiation the marginal cost curve is

$$MC = \frac{dTC}{dx} = 6x + 20$$

That is, by the process of differentiation the total cost function becomes the marginal cost function. If we are now told that a marginal cost function is

$$MC = 6x + 20$$

and asked to determine the total cost function we must reverse the process of differentiation, that is, we must *integrate* the marginal cost equation. From our knowledge of differential calculus we might guess that

$$TC = 3x^2 + 20x$$

and we should also realise that there may have been a constant term in the TC function which would have disappeared on differentiation. However, we cannot determine the value of this constant, if indeed there was one, from the marginal cost function alone. For example

$$TC = 3x^2 + 20x, \qquad TC = 3x^2 + 20x + 13, \qquad TC = 3x^2 + 20x + 50$$

each have the same marginal cost function

$$MC = 6x + 20$$

We therefore write the total cost function associated with this marginal cost function as

$$TC = 3x^2 + 20x + C$$

where C is an arbitrary constant, *the constant of integration*, whose value can be determined only if additional information is available. For example, if we know that $TC = 100$ when $x = 3$, then $C = 13$.

The notation commonly adopted is

$$\int (6x + 20)\, dx = 3x^2 + 20x + C$$

where the $\int$ is the *integral sign*, the process is known as *integration* and the result is called the *indefinite integral* of $6x + 20$. The term dx is included to show that x is the variable being considered.

By making use of the fact that integration can be thought of as the reverse of differentiation it is possible to find, by trial and error, the integral of many common functions. However, this may not always be easy because a function may be altered in appearance after differentiation by grouping together like terms or by cancelling out of common factors.

It is because of this that integration is generally more difficult than differentiation and it requires a good deal of practice and a bit of intuition to become more proficient. We have at our disposal, however, a number of methods for finding the integral of most common functions, and a number of examples are considered here to illustrate the method of approach.

6.2 EXERCISES

1. Derive the total cost function for processes in which the

   (a) marginal cost is £1 per unit and the fixed cost is £600

   (b) marginal cost is £1 per unit and the cost of production of 250 units is £300

   (c) marginal cost is given by

   $$MC = 2x + 3$$

   and the total cost is 100 when $x = 5$.

6.3 TECHNIQUES OF INTEGRATION

In this section the integrals of the functions discussed in the previous chapter are to be determined mainly by reversing the process of differentiation. More difficult forms are deferred until Section 6.10.

(a) If $y = ax^n$, $dy/dx = anx^{n-1}$

Hence,

$$\int ax^n \, dx = \frac{ax^{n+1}}{n+1} + C$$

which is true except in the case of $n = -1$ where the denominator would become zero. This exception is considered under (b) below.

Examples

1. $a = 1$, $n = 3$, $\int x^3 \, dx = \dfrac{x^4}{4} + C$

2. $a = 2$, $n = 0.5$, $\int 2x^{0.5} \, dx = \dfrac{2x^{1.5}}{1.5} + C$

3. $a = 4$, $n = -2$, $\int 4x^{-2} \, dx = \dfrac{4x^{-1}}{-1} + C$

(b) If $y = \log x$, $dy/dx = 1/x$

Hence,

$$\int \frac{1}{x} \, dx = \log x + C$$

which is the special case referred to above. More generally, if

$$y = \log u, \qquad \frac{dy}{dx} = \frac{1}{u}\frac{du}{dx}$$

Hence,

$$\int \frac{1}{u}\frac{du}{dx} = \log u + C$$

Thus if the numerator is the derivative of the denominator the integral is the logarithm of the denominator.

Examples

1. $u = x$, $\dfrac{du}{dx} = 1$ and $\int \dfrac{1}{x} \, dx = \log x + C$

155

2. $u = 2x + 3$, $\dfrac{du}{dx} = 2$ and $\displaystyle\int \dfrac{2\,dx}{2x+3} = \log\,(2x+3) + C$

3. $u = x^3 - 4x + 1$, $\dfrac{du}{dx} = 3x^2 - 4$

$$\int \frac{3x^2 - 4}{x^3 - 4x + 1}\,dx = \log\,(x^3 - 4x + 1) + C$$

4. In some cases the expressions to be integrated need to be modified slightly to obtain the standard form. For example,

$$\int \frac{x^2 + 2x}{x^3 + 3x^2 + 5}\,dx$$

Here, $u = x^3 + 3x^2 + 5$ and

$$\frac{du}{dx} = 3x^2 + 6x = 3(x^2 + 2x)$$

Therefore write

$$\frac{x^2 + 2x}{x^3 + 3x^2 + 5} = \frac{1}{3}\left(\frac{3x^2 + 6x}{x^3 + 3x^2 + 5}\right)$$

Hence

$$\int \frac{x^2 + 2x}{x^3 + 3x^2 + 5}\,dx = \frac{1}{3}\int \frac{3x^2 + 6x}{x^3 + 3x^2 + 5}\,dx = \tfrac{1}{3}\,\log\,(x^3 + 3x^2 + 5) + C$$

(c) If $y = ae^{bx}$, $dy/dx = abe^{bx}$

Hence $\displaystyle\int ae^{bx}\,dx = \frac{ae^{bx}}{b} + C$

Examples

1. $a = 3$, $b = 2$, $\displaystyle\int 3e^{2x}\,dx = \frac{3e^{2x}}{2} + C$

2. $a = 4$, $b = -1$, $\displaystyle\int 4e^{-x}\,dx = -4e^{-x} + C$

(d) If $y = \sin x$, $dy/dx = \cos x$

Hence $\qquad\qquad \int \cos x \, dx = \sin x + C$

Similarly $\qquad\qquad \int \sin x \, dx = -\cos x + C$

The more general versions of these and also the integrals of other trigonometric functions are deferred to Appendix A, Section 6.

6.4 EXERCISES

1. Obtain the total cost functions from the following marginal cost functions:

(a) $MC = 2x + 3$ with $TC = 50$ when $x = 5$

(b) $MC = x^2 + 2x + 4$ with $TC = 450$ when $x = 9$.

2. Integrate the following functions with respect to x:

(a) $3x - 4$ $\qquad\qquad\qquad$ (b) $2x^2 - 4x + 3$

(c) $3x^5 - \dfrac{4}{x^2} + 2x$ $\qquad\qquad$ (d) $\dfrac{2}{x^3} - \dfrac{4}{x}$

(e) $\dfrac{3}{3x - 5}$ $\qquad\qquad\qquad$ (f) $\dfrac{4x + 1}{2x^2 + x - 4}$

(g) $\dfrac{x^3 - x}{x^4 - 2x^2 + 2}$ $\qquad\qquad$ (h) $\dfrac{4x + 6}{x^2 + 3x + 1}$

(i) $2e^{3x} + e^{-x}$ $\qquad\qquad\quad$ (j) $2e^x - 4e^{3x}$

(k) $2 \sin x - 4 \cos x$

6.5 THE CALCULATION OF AREAS

The area under a curve can generally be determined by using the integral calculus.

Let AB be a section of the curve $y = f(x)$ between the values $x = X_1$, and $x = X_2$, and let Z be ABX_2X_1, the area which is required (Fig 6.1). In this illustration $f(x)$ is taken to be steadily increasing: the treatment in other cases is similar.

Let us consider two points P_1 and P_2 on the curve which are a small distance apart and from which perpendiculars are dropped to the x-axis. If the heights of these perpendiculars are y and $y + \delta y$

157

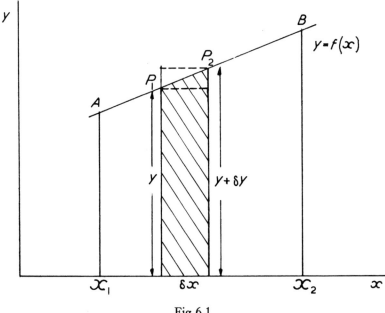

Fig 6.1

respectively and their horizontal distance apart is δx, then it is possible to estimate the area between these two lines, the curve, and the x-axis. This area, which we call δz, is shaded in the diagram and it must be greater than the area of the rectangle $y \, \delta x$ and less than the rectangle $(y + \delta y) \, \delta x$

that is

$$y \, \delta x < \delta z < (y + \delta y) \, \delta x$$

or

$$y < \frac{\delta z}{\delta x} < y + \delta y$$

When $\delta x \to 0$, $\delta y \to 0$ and $\delta z / \delta x \to dz/dx$, so that

$$\frac{dz}{dx} = y = f(x) \tag{1}$$

and the rate of change of the area depends on the value of x. The total area ABX_2X_1 is made up of an infinitely large number of these

158

small areas such as dz. Its value is therefore given by the sum of all such areas which are located between the values $x = X_1$, and $x = X_2$. This is usually written as

$$Z = \int_{x = X_1}^{x = X_2} \left(\frac{dz}{dx}\right) dx = \int_{X_1}^{X_2} f(x)\, dx \qquad (2)$$

where the elongated S is used as a summation sign.

This is known as a *definite integral* and it has a single numerical value associated with it. In this example the value corresponds to the area under the curve. The definite integral is obtained via the indefinite integral in the following way:

Step 1. Obtain the indefinite integral of the function by one of the methods discussed previously.

Step 2. Substitute the value $x = X_1$ in the indefinite integral.

Step 3. Substitute the value $x = X_2$ in the indefinite integral.

Step 4. Subtract the numerical value obtained in Step 2 from the numerical value obtained in Step 3 and the result is the value of the definite integral of the function between the limits $x = X_1$ and $x = X_2$.

A derivation of these rules is rather difficult but a justification is provided by the following simple example.

Example Let us consider the linear function $y = x$. This is a straight line which passes through the origin and bisects the angle between the x- and the y-axis (Fig 6.2).

Any point on this line has the same value for y as it does for x, and P, with co-ordinates (a, a), can be considered as a typical point.

The area which is bounded by the line $y = x$, the ordinate $x = a$ and the x-axis is triangular in shape and is shaded in the diagram. Its area is calculated very easily by simple geometry and is equal to $\frac{1}{2}a^2$. Let us compare this with the result that is obtained via the integral calculus.

Step 1. Determine the indefinite integral of the function with respect to x.

159

$$I = \int y \, dx$$

$$= \int x \, dx \text{ (by substitution } y = x)$$

$$= \tfrac{1}{2}x^2 + C$$

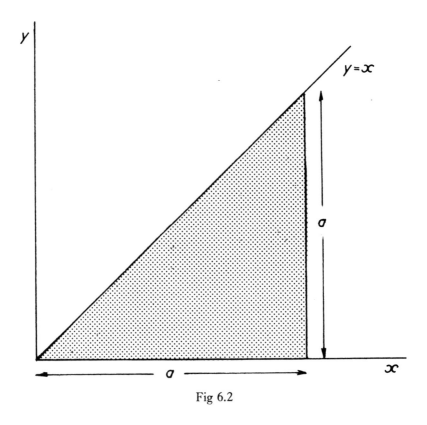

Fig 6.2

Step 2. Substitute the value $x = 0$ in the indefinite integral ($x = 0$ is the ordinate which corresponds to the left-hand side of the required area). The result is that

$$I_1 = 0 + C = C$$

Step 3. Substitute the value $x = a$ in the indefinite integral ($x = a$ is the ordinate which corresponds to the right-hand side of the required area). Then

$$I_2 = \tfrac{1}{2}a^2 + C$$

Step 4. Subtract I_1 from I_2

$$Z = I_2 - I_1$$
$$= \tfrac{1}{2}a^2 + C - C$$
$$= \tfrac{1}{2}a^2.$$

This is an identical result to that obtained by simple geometry and it is obvious now that the definite integral will never contain a constant of integration. This is always eliminated during Step 4 of the above method.

It is not really necessary to use the integral calculus to determine the areas in simple cases such as those where the answer is obvious from the geometry of the system. It can be of great use, however, in many other cases, particularly where the function is non-linear as the following example shows.

Example For a company using a process where the marginal cost of production is a function of the level of output, decreasing at first and then increasing when the output is above a certain level, the following might apply:

$$MC = 10 - q + \tfrac{1}{20}q^2, \qquad 0 < q \leqslant 20$$

This can be presented graphically as shown in Fig 6.3.

The area under the marginal cost curve between the values $q = 10$ and $q = 20$ is equal to the extra cost which must be incurred when production is increased from 10 units to 20 units of output.

Extra cost $= \displaystyle\int_{q=10}^{q=20} (MC)\, dq$

$$= \int_{10}^{20} (10 - q + \tfrac{1}{20}q^2)\, dq$$

$$= [10q - \tfrac{1}{2}q^2 + \tfrac{1}{60}q^3]_{10}^{20}$$

$$= \{10(20) - \tfrac{1}{2}(20)^2 + \tfrac{1}{60}(20)^3\} - \{10(10) - \tfrac{1}{2}(10)^2 + \tfrac{1}{60}(10)^3\}$$

$$= 200 - 200 + \frac{400}{3} - 100 + 50 - \frac{50}{3} = \frac{200}{3}$$

161

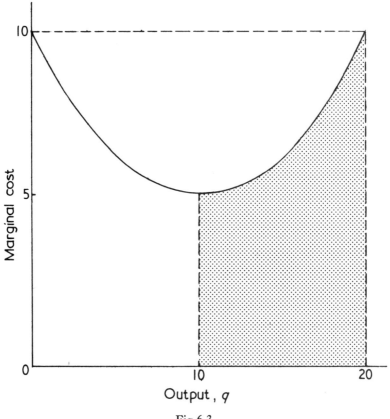

Fig 6.3

6.6   EXERCISES

1. Calculate the area between the axis of x, the straight line given
    by the following equations and the two values of x.

    (a) $y = 2x$ $x_1 = 0$ $x_2 = 3$

    (b) $y = 2x$ $x_1 = 3$ $x_2 = 6$

    (c) $y = 2x$ $x_1 = 0$ $x_2 = 6$

    (d) $y = 5x$ $x_1 = 6$ $x_2 = 8$

    (e) $y = 2 + 3x$ $x_1 = 0$ $x_2 = 3$

6.7 ABSOLUTE AREA

When the graph of $f(x)$ cuts the x-axis, areas below the x-axis are taken to be negative. For some applications this is the correct interpretation. For others, however, the *absolute area* under the curve is required, ie the deviations of the curve from the x-axis are to be summed. When this happens, the direct evaluation of the integral gives an incorrect solution. This can be seen by reference to the function $y = x$ and by considering the area shaded in Fig 6.4.

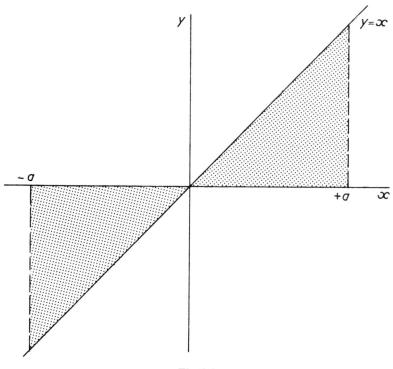

Fig 6.4

The shaded area between the ordinates $x = -a$ and $x = +a$ is obviously equal to $\frac{1}{2}a^2 + \frac{1}{2}a^2$, that is, a^2. Using the integral calculus as before would lead to

163

$$Z= \int\limits_{x=-a}^{x=+a} y\,dx = \int\limits_{x=-a}^{x=+a} x\,dx = [\tfrac{1}{2}x^2]_{-a}^{+a} = \tfrac{1}{2}a^2 - \tfrac{1}{2}a^2 = 0$$

This is true if we consider the area above the x-axis to be positive and the area below the x-axis to be negative. But if the total shaded area is required irrespective of whether it is above or below the axis, then the two sections must be calculated separately and their individual values added together. The dividing point is that at which the function crosses the x-axis. (Functions of higher order than the simple linear one may cross the axis in a number of points and in such cases it may be necessary to consider all the areas separately.)

In the example with the linear function $y=x$ the dividing point is at $x=0$ and the area is found as follows:

$$Z_1 = \int\limits_{x=-a}^{x=0} x\,dx = [\tfrac{1}{2}x^2]_{x=-a}^{x=0} = -\tfrac{1}{2}a^2$$

$$Z_2 = \int\limits_{x=0}^{x=a} x\,dx = [\tfrac{1}{2}x^2]_{x=0}^{x=a} = \tfrac{1}{2}a^2$$

The area is the sum of the absolute values of Z_1 and Z_2, ie the sum of their numerical values when the negative sign associated with Z_1 is ignored:

$$Z = \tfrac{1}{2}a^2 + \tfrac{1}{2}a^2 = a^2$$

6.8 EXERCISES

1. Sketch the following curves and calculate the absolute area between the curves, the axis of x and the given values x_1 and x_2.

    (a) $y=2x$ $x_1 = -1$ $x_2 = +1$

    (b) $y=2+3x$ $x_1 = -3$ $x_2 = +2$

    (c) $y=x^2$ $x_1 = -2$ $x_2 = +3$

    (d) $y=1+x+x^2$ $x_1 = -4$ $x_2 = +1$

6.9 CONSUMER'S SURPLUS

If the demand curve for a product is downward sloping, there will be some consumers who are willing to pay more than the current

market price in order to obtain the product. The gap between the
market price and the price the consumer is willing to pay is a measure
of the consumer's surplus satisfaction and is known as *consumer's
surplus*. Since the demand function is continuous the total consumer's
surplus is measured by the area ABP_1, shown in Fig 6.5, where P_1
and X_1 are the market price and quantity. This area can be evaluated
by integration as the area ABX_1O minus the area P_1BX_1O.

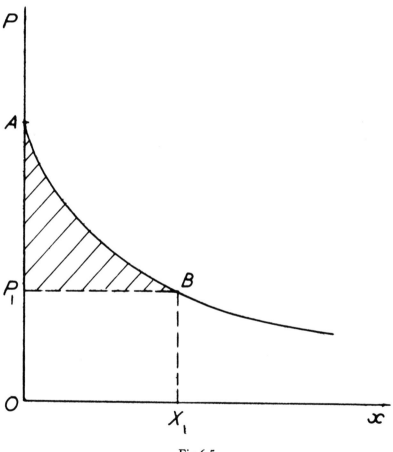

Fig 6.5

For example, if the demand function is

$$p = 45 - 2x - x^2$$

and the market price is $P_1 = 10$ and the quantity is $X_1 = 5$, then the area ABX_1O is given by

$$I = \int_{x=0}^{x=5} p \, dx = \int_{x=0}^{x=5} (45 - 2x - x^2) \, dx$$

$$= \left[45x - x^2 - \frac{x^3}{3} \right]_0^5$$

$$= \left(225 - 25 - \frac{125}{3} \right) - (0) = 158.33$$

The area P_1BX_1O is $P_1X_1 = 50$ and therefore the consumer's surplus is measured as $158.33 - 50 = 108.33$.

6.10 MORE DIFFICULT INTEGRATION

INTEGRATION BY PARTS

Integration by parts can be thought of as the reverse of differentiation of a product. We know that if u and v are functions of x then

$$\frac{d}{dx}(uv) = u \frac{dv}{dx} + v \frac{du}{dx}$$

Integrating both sides of this, we have

$$\int \frac{d}{dx}(uv) \, dx = \int u \frac{dv}{dx} \, dx + \int v \frac{du}{dx} \, dx$$

$$uv = \int u \, dv + \int v \, du$$

$$\therefore \quad \int u \, dv = uv - \int v \, du$$

Thus to find the integral of a function such as xe^x we can let $x = u$ and $e^x \, dx = dv$.

Then $$v = \int dv = \int e^x \, dx = e^x, \quad \text{and} \quad du = dx$$

$$\therefore \quad \int xe^x \, dx = xe^x - \int e^x \, dx$$

$$= xe^x - e^x + C$$

$$= e^x(x - 1) + C$$

We can check this by differentiation:

$$\frac{d}{dx}\left[e^x(x-1)+C\right]=e^x(x-1)+e^x=xe^x$$

This method is of great importance in integration but care must be taken in deciding which part of the function should be represented by u and which by v. It is usually better to represent by dv that part which can be integrated most easily. If both parts are simple to integrate then represent by u that part which is reduced to a constant value by differentiation in the shortest possible time. There is, however, no hard and fast rule about this selection. The following examples have been chosen to show the method of selection in a few typical cases.

Example 1

$$\int x^2 e^x \, dx$$

Let $\qquad\qquad x^2 = u \qquad$ and $\qquad e^x \, dx = dv$

Then $\qquad\qquad \int x^2 e^x \, dx = x^2 e^x - \int e^x 2x \, dx$

$$= x^2 e^x - 2\int e^x x \, dx$$

The second part of this must be integrated by the same method

Let $\qquad\qquad x = u \qquad$ and $\qquad e^x \, dx = dv$

Then $\qquad\qquad \int e^x x \, dx = \int xe^x \, dx$

$$= xe^x - \int e^x \, dx$$

$$= xe^x - e^x + C$$

$$= e^x(x-1) + C$$

Then taking the two parts together, we have

$$\int x^2 e^x \, dx = x^2 e^x - 2e^x(x-1) + C$$

$$= e^x(x^2 - 2x + 2) + C$$

It is not necessary to represent the terms by u and v in the order in which they appear in the function as the second part of Example 1 and Example 2 show.

Example 2

$$\int x \log_e x \, dx$$

Let $\qquad\qquad \log_e x = u \qquad$ and $\qquad x \, dx = dv$

Then $\qquad\qquad \dfrac{1}{x} \, dx = du \qquad$ and $\qquad \frac{1}{2}x^2 = v + C$

$\therefore \quad \int x \log_e x \, dx = \int (\log_e x) x \, dx$

$$= (\log_e x)(\tfrac{1}{2}x^2) - \int (\tfrac{1}{2}x^2) \left(\frac{1}{x}\right) dx$$

$$= \tfrac{1}{2}x^2 \log_e x - \tfrac{1}{2} \int x \, dx$$

$$= \tfrac{1}{2}x^2 \log_e x - \tfrac{1}{4}x^2 + C$$

$$= \tfrac{1}{2}x^2(\log_e x - \tfrac{1}{2}) + C$$

It may be necessary to use more than two stages to complete the integration, but when this happens we should be reasonably certain that we are applying the appropriate technique. In general, this method reduces the part to be integrated to a simpler form at each stage.

THE METHOD OF PARTIAL FRACTIONS

In attempting to integrate some functions it is convenient to express the function as a series of partial fractions. For example, to integrate $(a^2 - x^2)^{-1}$ notice that

$$a^2 - x^2 = (a + x)(a - x)$$

and hence

$$\frac{1}{a^2 - x^2} = \frac{1}{(a + x)(a - x)} = \frac{A}{a + x} + \frac{B}{a - x}$$

where A and B are as yet unknown constants. Their values are determined as follows:

$$\frac{1}{a^2 - x^2} = \frac{A}{a + x} + \frac{B}{a - x} = \frac{A(a - x) + B(a + x)}{a^2 - x^2}$$

Since the denominators are equal then

$$1 = A(a - x) + B(a + x)$$
$$= (Aa + Ba) + x(B - A)$$

This can be true only if the coefficients of x in this equation are equal, so that

$$0 = B - A$$

and also the constant terms are equal so that

$$1 = Aa + Ba$$

From these two equations the values of A and B which satisfy the identity can be found and the result is

$$A = \frac{1}{2a}, \qquad B = \frac{1}{2a}$$

The original problem can now be restated as

$$\int \frac{1}{(a^2 - x^2)} \, dx = \int \left[\frac{1/2a}{(a + x)} + \frac{1/2a}{(a - x)} \right] dx$$

$$= \frac{1}{2a} \left[\int \frac{1}{(a + x)} \, dx + \int \frac{1}{(a - x)} \, dx \right]$$

$$= \frac{1}{2a} [\log_e (a + x) - \log_e (a - x)] + C$$

The negative sign in front of the second term is due to the fact that the denominator of the second fraction has a negative sign prefixing the variable x. Now the difference between the logarithms of two numbers can be written as the logarithm of the quotient of the numbers:

$$\int \frac{1}{(a^2 - x^2)} \, dx = \frac{1}{2a} \log_e \left(\frac{a + x}{a - x} \right) + C$$

The method of partial fractions can be extended to cover cases where the denominator is of higher degree than the second. It may be possible to split the denominator into several factors but this is likely to result in a factor which is a quadratic. In this case we let the numerator of the fraction containing this factor be linear in x. For example,

169

$$\int \frac{1}{(x+2)(x^2+3x+1)}\,dx \equiv \int \left[\frac{A}{(x+2)} + \frac{Bx+C}{(x^2+3x+1)} \right] dx$$

where
$$1 \equiv A(x^2+3x+1)+(Bx+C)(x+2)$$
$$\equiv Ax^2+3Ax+A+Bx^2+2Bx+Cx+2C$$
$$\equiv x^2(A+B)+x(3A+2B+C)+(A+2C)$$

Then equating the coefficients of x^2, of x and of the constant term on each side of the identity we obtain three equations in three unknowns.

$$A+B=0 \tag{1}$$

$$3A+2B+C=0 \tag{2}$$

$$A+2C=1 \tag{3}$$

Solving these equations gives

$$A=-1, \qquad B=1, \qquad C=1.$$

$$\therefore \quad \frac{1}{(x+2)(x^2+3x+1)} = \frac{-1}{x+2} + \frac{x+1}{x^2+3x+1}$$

The first term on the right-hand side is integrated very easily:

$$\int \frac{-1}{x+2}\,dx = -\log_e(x+2)$$

The second term on the right-hand side requires further consideration. It can be written as

$$\frac{x+1}{x^2+3x+1} = \frac{x+\frac{3}{2}-\frac{1}{2}}{x^2+3x+1}$$

$$= \frac{\frac{1}{2}(2x+3-1)}{x^2+3x+1}$$

$$= \frac{1}{2}\left[\frac{2x+3}{x^2+3x+1} - \frac{1}{x^2+3x+1} \right]$$

Again, the first term on the right-hand side can be integrated very easily because the numerator is the derivative of the denominator.

170

$$\int \frac{1}{2}\left(\frac{2x+3}{x^2+3x+1}\right) dx = \tfrac{1}{2}\log_e(x^2+3x+1)$$

The only term remaining is

$$\int \frac{-1}{x^2+3x+1} dx$$

Since $-(x^2+3x+1) = (\tfrac{5}{4})-(x+\tfrac{3}{2})^2$ and, from above,

$$\int \frac{1}{a^2-x^2} dx = \frac{1}{2a}\log_e\left(\frac{a+x}{a-x}\right)+C$$

then $$\int \frac{1}{\frac{5}{4}-(x+\frac{3}{2})^2} dx = \frac{2}{2\sqrt{5}}\log_e\left(\frac{\sqrt{\frac{5}{4}}+x+\frac{3}{2}}{\sqrt{\frac{5}{4}}-x-\frac{3}{2}}\right)+C$$

Collecting the three parts together gives

$$\int \frac{1}{(x+2)(x^2+3x+1)} dx = -\log_e(x+2)+\tfrac{1}{2}\log_e(x^2+3x+1)$$
$$+\frac{1}{2\sqrt{5}}\log_e\left(\frac{\sqrt{\frac{5}{4}}+x+\frac{3}{2}}{\sqrt{\frac{5}{4}}-x-\frac{3}{2}}\right)+C$$

The result could be further simplified but this is left to the reader. The method may appear rather complicated but it serves to illustrate the way in which the integral of more complicated functions can often be obtained by the breaking down of the function in successive stages into a series of terms for which the integral is known.

The major problem is to recognise the most suitable way in which to break the function down in order to arrive at a standard form. Care must then be taken in the application of such techniques as partial fractions to ensure that the function which is finally integrated is in fact identical in all respects with the original function. It is important that the reader should consult a more advanced treatment of the integral calculus before tackling such problems, as a number of quite simple, but important, modifications to the above procedure may be necessary in different cases.

171

INTEGRATION BY SUBSTITUTION

This can be illustrated by a simple example where the result can be recognised at a glance.

Example 1

$$\int \frac{2x}{x^2+2}\, dx$$

Let

$$u = x^2 + 2$$

Then

$$\frac{du}{dx} = 2x$$

$\therefore$

$$du = 2x\, dx$$

Substituting these values in the original expression, we have

$$\int \frac{2x}{x^2+2}\, dx = \int \frac{1}{u}\, du = \log_e u + C$$

Then substituting back the value (x^2+2) for u we have

$$\int \frac{2x}{x^2+2}\, dx = \log_e (x^2+2) + C$$

This result was, of course, obvious. A more realistic example is the following.

Example 2

$$\int (4x-3)^6\, dx$$

This could be obtained by use of the binomial expansion. However,

let $\qquad u = 4x - 3 \qquad$ so that $\qquad du = 4\, dx$

Then

$$\int (4x-3)^6\, dx = \int \frac{u^6}{4}\, du = \frac{u^7}{28} + C$$

$$= \frac{(4x-3)^7}{28} + C$$

Other examples of integration by substitution are provided in Appendix A, Sections 6 and 7, where trigonometric substitutions are used. Some standard integrals considered in this chapter are given in Table 6.1.

TABLE 6.1

Some standard integrals
(The constants of integration are omitted.)

Function	Integral	Conditions
ax^n	$\dfrac{ax^{n+1}}{n+1}$	$n \neq -1$
$\dfrac{1}{x}$	$\log_e x$	
$\dfrac{1}{u}\dfrac{du}{dx}$	$\log_e u$	u is a function of x
$a\,e^{bx}$	$\dfrac{a\,e^{bx}}{b}$	
$\cos x$	$\sin x$	
$\sin x$	$-\cos x$	
$u\,dv$	$uv - \int v\,du$	u, v are functions of x

6.11 EXERCISES

Integrate the following expressions with respect to x.

1. $(5x^2 + 4)e^{3x}$

2. $x^2 \log_e x$

3. $\dfrac{3}{(x+1)(x-1)}$

4. $\dfrac{x}{(x-2)(x^2-3)}$

5. $(4x+3)^{10}$

6. $\log_e x$

CHAPTER SEVEN

Partial Differentiation

7.1 FUNCTIONS OF MORE THAN ONE VARIABLE

Our concern so far has been with the relationship between two variables and it is obvious that little progress can be made if we continue to confine ourselves in this way, because the majority of economic relationships involve more than two variables. For example, the demand for a good depends not only on its price, but on the prices of other goods, and on the consumers' income. We must, therefore, expect to deal with functions such as

$$q = f(p_1, p_2, I, \ldots)$$

and become familiar with methods of analysis which enable us to handle such situations.

These methods are a logical extension of the differential calculus, but to avoid confusion a new and slightly different notation is required. To introduce this notation and to illustrate its use, let us consider a three-variable model which it is possible to represent in diagrammatic form.

Let x and y be two independent variables and z the dependent variable. Then $z = f(x, y)$ and if we represent x and y on axes at right angles to each other in the horizontal plane we can imagine z as being on an axis in the vertical plane and, therefore, at right angles to both x and y. Any variation in x or y then causes a change in z and the point representing z moves, in fact, in such a way as to trace out a surface above the plane xy rather like a hill or series of hills on the earth's surface.

The point P in Fig 7.1 has a height above the xy-plane equal to z, and therefore its z-co-ordinate is given by $z = z_1$. To completely

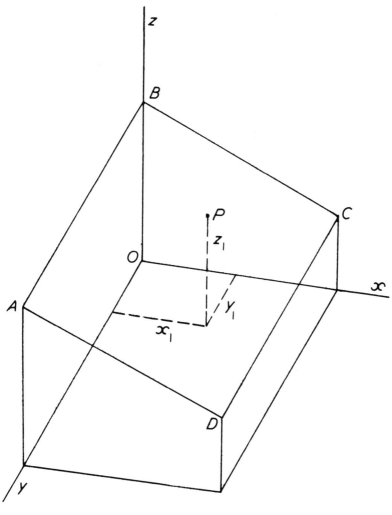

Fig 7.1

determine the location of P in the three-dimensional space it is also necessary to know the distance of the point from the yz-plane and from the xz-plane. These distances are $x = x_1$ and $y = y_1$ respectively and the point P can then be specified uniquely in relation to the three axes by the co-ordinates (x_1, y_1, z_1).

Because $z = f(x, y)$ the point P moves over the surface $ABCD$ as the values of x and y change, and we are interested in measuring the rate of change of z with respect to changes in the other variables. In fact, what is required is some value similar to the derivative, which can be used in the three-variable case to measure rates of change.

This gives us a basis on which to work. If one of the variables, say y, is considered to be fixed, then this is equivalent to taking a section through the model parallel to the x-axis as shown in Fig 7.2. Any change in z on this section is then due entirely to the change in x.

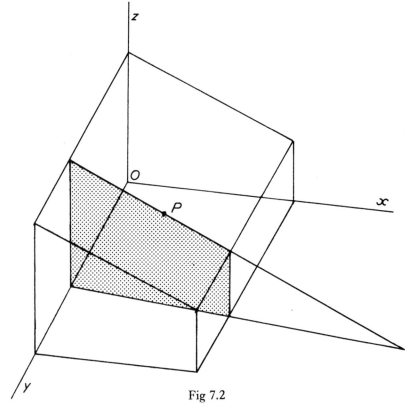

Fig 7.2

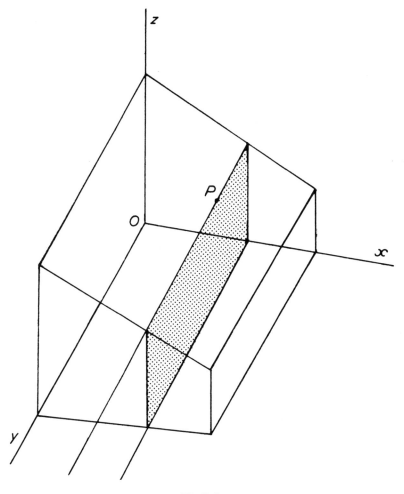

Fig 7.3

Any point P on this section is, therefore, similar to the point in a two-dimensional model. A tangent can be drawn to the surface at P parallel to the x-axis and the slope of this line is a measure of the rate of change of z with respect to x when y is kept constant. It is therefore a form of derivative, but because it is the rate of change with respect to only one variable, the other variable being held constant, it is called a partial derivative, and to prevent confusion with dz/dx it is denoted by $\partial z/\partial x$ or z_x.

In a similar manner it is possible to take a section through the solid for a constant value of x and determine the rate of change of z as y changes. This ratio is written $\partial z/\partial y$ or z_y and is a measure of the slope of the section shown in Fig 7.3.

It is not easy to obtain the value of the derivative from a diagram and, as might be expected, it is not necessary because with only minor modification, the differential calculus can be used to obtain the result very simply. The modification is necessary because z_x refers to the partial derivative of z with respect to the variable x and implies that all other variables in the function must be considered as constant. For example, if

$$z = x^2 + y^2$$

$$\frac{\partial z}{\partial x} = 2x$$

This follows because y^2 is considered to be a constant and the derivative of a constant is zero. Similarly,

$$\frac{\partial z}{\partial y} = 2y$$

because x^2 is considered as a constant when the partial derivative with respect to y is required.

The same procedure applies if the expression is more complicated. For example,

$$z = x^2 + 3xy + y^2$$

To obtain the partial derivatives of z with respect to x only, we treat y as though it were a constant wherever it appears.

$$\frac{\partial z}{\partial x} = 2x + 3y$$

Similarly
$$\frac{\partial z}{\partial y} = 3x + 2y$$

The second-order partial derivatives are defined in a similar way to second-order derivatives, ie they are the result of differentiating the first-order derivatives. We define

$$z_{xx} = \frac{\partial^2 z}{\partial x^2} = \frac{\partial}{\partial x}(z_x)$$

$$z_{yy} = \frac{\partial^2 z}{\partial y^2} = \frac{\partial}{\partial y}(z_y)$$

$$z_{yx} = \frac{\partial^2 z}{\partial y \partial x} = \frac{\partial}{\partial y}\left(\frac{\partial z}{\partial x}\right) = \frac{\partial}{\partial y}(z_x)$$

$$z_{xy} = \frac{\partial^2 z}{\partial x \partial y} = \frac{\partial}{\partial x}\left(\frac{\partial z}{\partial y}\right) = \frac{\partial}{\partial x}(z_y)$$

Here there are four second-order derivatives, z_{xx} and z_{yy} which correspond to d^2y/dx^2 and also the two 'mixed' derivatives z_{xy} and z_{yx}. The notation used indicates the order in which the variables occur, ie z_{yx} is found by differentiating z with respect to x and then with respect to y. In most cases

$$z_{xy} = z_{yx}$$

and the order of differentiation does not matter,
For example $z = x^2 + 3xy + 2y^3 - 5x - 4y + 20$

$$z_x = \frac{\partial z}{\partial x} = 2x + 3y - 5$$

$$z_y = \frac{\partial z}{\partial y} = 3x + 6y^2 - 4$$

$$z_{xx} = \frac{\partial}{\partial x}(z_x) = 2$$

$$z_{yy} = \frac{\partial}{\partial y}(z_y) = 12y$$

$$z_{yx} = \frac{\partial}{\partial y}(z_x) = 3$$

179

$$z_{xy} = \frac{\partial}{\partial x}(z_y) = 3$$

and hence $\qquad z_{xy} = z_{yx}$

The above definitions of first- and second-order partial derivatives can easily be generalised to cover the cases of more variables and higher-order derivatives.

7.2 MARGINAL ANALYSIS

This is an extension of the two-dimensional model and is best explained by the use of an example.

Let us assume a function in which the level of output is related to the quantity of inputs x and y in the following way:

$$q = f(x, y) = 50x - x^2 + 60y - 2y^2$$

then
$$\frac{\partial q}{\partial x} = 50 - 2x$$

$$\frac{\partial q}{\partial y} = 60 - 4y$$

These partial derivatives denote the marginal productivity of factor x and factor y respectively, and in this example they are a function of the amount already in use. That is, if we are at present using 10 units of each factor then

$$\frac{\partial q}{\partial x} = 50 - 2 \times 10 = 30$$

$$\frac{\partial q}{\partial y} = 60 - 4 \times 10 = 20$$

and the marginal productivity of x is 30 whilst that of y is 20.

An example which is more useful in economics is the *Cobb–Douglas production function*

$$q = AL^\alpha C^\beta$$

where $\qquad q = \text{output}$

180

$$C = \text{capital stock}$$

$$L = \text{labour input}$$

A, α, β are constants. The marginal product of labour is

$$q_L = \frac{\partial q}{\partial L} = \alpha A L^{\alpha-1} C^\beta = \frac{\alpha A L^\alpha C^\beta}{L} = \frac{\alpha q}{L}$$

and the marginal product of capital is

$$q_C = \frac{\partial q}{\partial C} = \beta A L^\alpha C^{\beta-1} = \frac{\beta q}{C}$$

It can easily be seen that these marginal products are positive if $\alpha > 0$ and $\beta > 0$, since q, C, L are all positive.

The way in which these marginal products are changing can be seen from the second-order derivatives

$$q_{LL} = \frac{\partial^2 q}{\partial L^2} = \alpha \left(\frac{L q_L - q}{L^2} \right) = \frac{\alpha(\alpha-1)q}{L^2}$$

$$q_{CC} = \frac{\partial^2 q}{\partial C^2} = \beta \left(\frac{C q_C - q}{C^2} \right) = \frac{\beta(\beta-1)q}{C^2}$$

It is usually a requirement of production functions that $q_{LL} < 0$ and $q_{CC} < 0$, that is that the marginal product is declining, and this will be the case if $1 > \alpha > 0$ and $1 > \beta > 0$.

7.3 ELASTICITY OF DEMAND

If the quantity demanded of a good (A) is determined by the price of that good and also the price of a substitute (B) then the demand function can be written as

$$q_D = f(p_A p_B)$$

The following ratios are then important:

1. The change in the quantity demanded of A when the price of A is changed and the price of B is held constant. This ratio can be expressed in partial-derivative notation as

$$\frac{\partial q_D / q_D}{\partial p_A / p_A} \quad \text{or} \quad \frac{\partial q_D}{\partial p_A} \frac{p_A}{q_D}$$

181

It is referred to as the *partial elasticity of demand for good A with respect to the price of good A*, or the *own-price elasticity of demand for good A*.

For example, suppose that

$$q_D = f(p_A p_B) = 100 - 3p_A + p_B$$

then

$$\frac{\partial q_D}{\partial p_A} = -3$$

∴ partial elasticity of demand $= -3\dfrac{p_A}{q_D}$

If the present prices are $p_A = 10$, $p_B = 20$

$$q_D = 100 - 3(10) + 20 = 90$$

∴ partial elasticity of demand $= -3 \cdot \dfrac{10}{90} = -\tfrac{1}{3}$

This can be interpreted as meaning that the quantity demanded of A falls by $\tfrac{1}{3}$ per cent if its price is increased by 1 per cent and the price of B remains unchanged.

2. The change in the quantity demanded of A when the price of A is held constant but the price of B is changed. This ratio can be expressed in partial-derivative notation as:

$$\frac{\partial q_D / \partial p_B}{q_D / p_B} \quad \text{or} \quad \frac{\partial q_D}{\partial p_B} \frac{p_B}{q_D}$$

It is referred to as the *partial elasticity of demand for good A with respect to the price of good B* or alternatively as the *cross-elasticity of demand for A with respect to B*.

In the example previously used, when $p_B = 20$,

$$\frac{\partial q_D}{\partial p_B} = 1 \quad \text{and} \quad q_D = 90$$

∴ Cross-elasticity of demand $= 1 \times \dfrac{20}{90} = 0.22$

This means that the demand for A increases by 0.22 per cent if its price remains unchanged and the price of B is increased by 1 per cent.

An extension of the above analysis to include the cases where the dependent variable is a function of more than two independent variables is very simple. The partial derivative with respect to any one of the variables is obtained by differentiating the function with respect to this variable and treating all other variables as constants.

For example, suppose that

$$q_D = a_1 p_1 + a_2 p_2 + a_3 p_3 + a_4 p_4$$

then

$$\frac{\partial q_D}{\partial p_1} = a_1, \qquad \frac{\partial q_D}{\partial p_2} = a_2$$

$$\frac{\partial q_D}{\partial p_3} = a_3, \qquad \frac{\partial q_D}{\partial p_4} = a_4$$

7.4   EXERCISES

1. Find $\partial z/\partial x$ and $\partial z/\partial y$ for the following functions:

   (a) $z = xy + x^2 y + xy^2$

   (b) $z = 3x^2 y + 4xy^2 + 6xy$

   (c) $z = x^2 \sin y + y^2 \cos x$

   (d) $z = (x+y)e^{(x+y)}$

   (e) $z = \log (x^2 + y^2)$

   (f) $z = \dfrac{x+y}{x-y}$

   (g) $z = \dfrac{x + \sin y}{y + \sin x}$

2. Let the output of a product be given by

   $$q = 2(C-30)^2 + 3(L-20)^2 + 2(C-30)(L-20)$$

   where L and C are measures of capital and labour respectively. Show that the marginal products of capital and labour are equal when $C = 50$ and $L = 30$.

183

3. Let the demand for apples be given by

$$q = 240 - p_a^2 + 6p_o - p_a p_o$$

where q is the quantity of apples demanded at price p_a, and p_o is the price of oranges. Evaluate the partial elasticity of demand for apples when $p_a = 5$ and $p_o = 4$ with respect to

(a) the price of apples

(b) the price of oranges.

7.5 DIFFERENTIALS

In the two-variable model, $y = f(x)$, we obtained a value for the magnitude of the change in y corresponding to a small but finite change in x by using the expression

$$dy = \frac{dy}{dx} dx$$

This procedure can be extended when there are more variables in the model. In fact, for $z = f(x, y)$ the small change in z is equal to the small change in x multiplied by the rate of change of z with respect to x, keeping y constant, plus the small change in y multiplied by the rate of change of z with respect to y keeping x constant. This is conveniently written as

$$dz = \frac{\partial z}{\partial x} dx + \frac{\partial z}{\partial y} dy$$

This can be extended to any number of variables. For example,

suppose $\qquad z = f(p, q, r, s, \ldots)$

then $\qquad dz = \dfrac{\partial z}{\partial p} dp + \dfrac{\partial z}{\partial q} dq + \dfrac{\partial z}{\partial r} dr + \dfrac{\partial z}{\partial s} ds + \cdots$

This expression is known as the *total differential* of z.

This can be illustrated by reference to the production function relating the quantity of wheat produced to the quantity of labour (x) and land (y) which is employed.

$$q = 40x - x^2 + 60y - 2y^2$$

$$\frac{\partial q}{\partial x} = 40 - 2x$$

$$\frac{\partial q}{\partial y} = 60 - 4y$$

$$\therefore \qquad dq = \frac{\partial q}{\partial x} \, dx + \frac{\partial q}{\partial y} \, dy$$

$$= (40 - 2x) \, dx + (60 - 4y) \, dy$$

This equation can be used to obtain an approximate value for the increased production which is possible when both labour and land are increased together. The smaller the increases in the variables, dx and dy that are considered, the more accurate the answer will be. However, an approximate answer can be obtained when the supply of each factor is increased by 1 unit. If the present quantity in use is given by $x = 10$ units, $y = 10$ units, then the approximate increase will be given by

$$dq = [40 - 2(10)] + [60 - 4(10)]$$

$$= 20 + 20$$

$$= 40$$

The accurate value of the increase can be obtained by substituting $x = 11$, $y = 11$ in the production function to give $q = 737$ and subtracting the value of q when $x = 10$, $y = 10$. The accurate increase is 37 units.

Differentials are also useful with *indifference curves*. If the utility function is

$$u = f(q_1, q_2)$$

where q_1 and q_2 are the quantities of two goods consumed, then a given level of utility, u^0, can be achieved with different combinations of q_1 and q_2. The expression

$$u^0 = f(q_1, q_2)$$

185

gives the *indifference curve* and the total differential of u^0 is

$$du^0 = \frac{\partial f}{\partial q_1} dq_1 + \frac{\partial f}{\partial q_2} dq_2$$

Since u^0 is a constant (ie any movement is along the indifference curve) then $du^0 = 0$ and

$$0 = f_1 dq_1 + f_2 dq_2$$

or $$\frac{dq_2}{dq_1} = -\frac{f_1}{f_2} = -\frac{\text{marginal utility of good 1}}{\text{marginal utility of good 2}}$$

This ratio is the rate of commodity substitution and is the slope of the indifference curve.

The same procedure can be applied to the production function

$$Q = f(K, L)$$

with the result that

$$\frac{dK}{dL} = -\frac{\partial f/\partial L}{\partial f/\partial K} = -\frac{Q_L}{Q_K} = -r$$

that is, the marginal rate of substitution (r) of labour inputs for capital input equals the ratio of the marginal products of the inputs.

7.6 TOTAL DERIVATIVES

Many situations arise where the independent variables can all be expressed as a function of a single variable such as time. For example,

$$z = x^2 + y^2$$

with $$x = e^t \cos t, \qquad y = e^t \sin t$$

What is required is the change in z which is brought about by the change in t as the latter affects both x and y. This can be obtained by substituting the function of t for x and y in the main equation and then differentiating this expression with respect to t. The same result can be obtained in a simpler way by making use of the results of the section on differentials (5.10).

$$dz = \frac{\partial z}{\partial x} dx + \frac{\partial z}{\partial y} dy$$

Dividing through this expression by dt

$$\frac{dz}{dt} = \frac{\partial z}{\partial x}\frac{dx}{dt} + \frac{\partial z}{\partial y}\frac{dy}{dt}$$

This is similar to the function of a function or chain rule discussed in Section 5.2.

For example in the case where $z = x^2 + y^2$ and $x = e^t \cos t$, $y = e^t \sin t$, we can form the derivatives

$$\frac{dx}{dt} = e^t \cos t - e^t \sin t = e^t(\cos t - \sin t)$$

$$\frac{dy}{dt} = e^t \cos t + e^t \sin t = e^t(\cos t + \sin t)$$

$$\frac{dz}{dt} = \frac{\partial z}{\partial x}\frac{dx}{dt} + \frac{\partial z}{\partial y}\frac{dy}{dt}$$

$$= 2xe^t(\cos t - \sin t) + 2ye^t(\cos t + \sin t)$$

This can be re-arranged by gathering together terms containing $\cos t$ and terms containing $\sin t$ with the result

$$\frac{dz}{dt} = e^{2t}(2\cos^2 t - 2\sin t \cos t + 2\cos t \sin t + 2\sin^2 t) = 2e^{2t}$$

7.7 EXERCISES

1. The quantity of output of a product depends on the labour (L) and capital (C) inputs in such a way that

$$q = L^3 - 3L + 4LC - C^2$$

If $L = 10$, $C = 5$, what is the approximate increase in q for an increase of 1 in both L and C? What is the approximate increase in q if L alone increases by 1?

2. The demand for apples (a) is related to the price of oranges (o) by

$$q = 240 - p_a^2 + 6p_o - p_a p_o$$

What is the approximate change in the demand for apples if the price of apples increased from $p_a = 5$ to $p_a = 6$ and the price of oranges changes from $p_o = 4$ to $p_o = 5$?

187

3. Find dx/dt, dy/dt and dz/dt for the following functions:

(a) $z = x^2 - y^2$, $x = e^t \cos t$, $y = e^t \sin t$

(b) $z = \log_e (x + y)$, $x = t^2 e^t$, $y = \sin t$

(c) $z = y e^{x^2}$, $x = e^t$, $y = \log_e t$

(d) $z = x^2 + xy + y^2$, $x = t + \dfrac{1}{t}$, $y = t - \dfrac{1}{t}$

7.8 IMPLICIT DIFFERENTIATION

The functions we have met have all been explicit, that is one variable has been expressed as a function of one or more other variables. For example, $y = x^2 + 3x$ *is an explicit function*. In contrast, for $x^3 + 2x^2 y = 3xy^2 + y^3$ an implicit function of the form $y = f(x)$ is involved.

It often happens that we require the derivative of y with respect to x for such functions. No difficulty arises in the first example as dy/dx is simply $2x + 3$ but it is not easy to see how the derivative can be obtained in the second case.

However, we do know that if $z = f(x, y)$ then approximately

$$\delta z = \frac{\partial z}{\partial x} \delta x + \frac{\partial z}{\partial y} \delta y$$

Dividing this throughout by δx, we have approximately

$$\frac{\delta z}{\delta x} = \frac{\partial z}{\partial x} \frac{\delta x}{\delta x} + \frac{\partial z}{\partial y} \frac{\delta y}{\delta x}$$

and as $\delta x \to 0$ this becomes

$$\frac{dz}{dx} = \frac{\partial z}{\partial x} + \frac{\partial z}{\partial y} \frac{dy}{dx}$$

If we have a relation in which $z = 0$ (identically), then dz/dx also equals 0.

This fact can be used as follows:

Suppose $z = x^3 + 2x^2 y - 3xy^2 - y^3 = 0$

then
$$\frac{dz}{dx} = 0$$

$$\frac{\partial z}{\partial x} = 3x^2 + 4xy - 3y^2$$

$$\frac{\partial z}{\partial y} = 2x^2 - 6xy - 3y^2$$

Substituting these values in the total derivative

$$\frac{dz}{dx} = \frac{\partial z}{\partial x} + \frac{\partial z}{\partial y}\frac{dy}{dx}$$

results in

$$0 = (3x^2 + 4xy - 3y^2) + (2x^2 - 6xy - 3y^2)\frac{dy}{dx}$$

and rearranging these terms, we have

$$\frac{dy}{dx} = -\frac{(3x^2 + 4xy - 3y^2)}{(2x^2 - 6xy - 3y^2)}$$

The above method can be generalised and used to determine the derivative of any implicit function of two variables. The function is first rearranged so that all the terms are on one side of the equality sign, ie it is made equal to 0. Then dz/dx is equal to zero and the following holds:

$$0 = \frac{\partial z}{\partial x} + \frac{\partial z}{\partial y}\frac{dy}{dx}$$

$$\therefore \qquad \frac{dy}{dx} = -\frac{\partial z/\partial x}{\partial z/\partial y}$$

For example, suppose

$$y^2 = 3xy + x^2$$

then
$$z = y^2 - 3xy - x^2 = 0$$

$$\frac{\partial z}{\partial y} = 2y - 3x$$

189

$$\frac{\partial z}{\partial x} = -3y - 2x = -(3y + 2x)$$

and

$$\frac{dy}{dx} = \frac{3y + 2x}{2y - 3x}$$

An alternative approach is to differentiate each term with respect to x, remembering that the function of a function rule must be used on such terms as y^2 and xy. For example, suppose

$$y^2 = 3xy + x^2$$

Differentiating term by term gives

$$2y\frac{dy}{dx} = 3x\frac{dy}{dx} + 3y + 2x$$

that is

$$\frac{dy}{dx}(2y - 3x) = 3y + 2x$$

or

$$\frac{dy}{dx} = \frac{3y + 2x}{2y - 3x}$$

Similarly:

$$0 = x^3 + 2x^2y - 3xy^2 - y^3$$

Differentiating gives

$$0 = 3x^2 + 2x^2\frac{dy}{dx} + 4xy - 6xy\frac{dy}{dx} - 3y^2 - 3y^2\frac{dy}{dx}$$

$$= 3x^2 + 4xy - 3y^2 + \frac{dy}{dx}(2x^2 - 6xy - 3y^2)$$

$$\therefore \quad \frac{dy}{dx} = \frac{3y^2 - 3x^2 - 4xy}{2x^2 - 6xy - 3y^2}$$

7.9   EXERCISES

1. Determine dy/dx if

   (a) $y^3 - 4x^2y^2 + 6x^2y + 4x = 0$

   (b) $x^2y^2 + e^x - e^y = 0$

   (c) $3\log_e(x + y) + 4x^2 - y^2 = 0$

   (d) $\sin(x + y) - 3xy = 10xy$

2. What is the elasticity of demand when $p = 2$ if

$$q^2 = \frac{500 - p^2}{2 + p} + pq$$

7.10 HOMOGENEOUS FUNCTIONS

A function is said to be *homogeneous of degree k* if

$$f(tx, ty) = t^k f(x, y)$$

for all values of t, where t and k are constants.

An example of a homogeneous function of degree one is

$$z = f(x, y) = 2x + 3y$$

Replacing x by tx and y by ty gives

$$f(tx, ty) = 2(tx) + 3(ty) = t(2x + 3y)$$

$$= tf(x, y) = tz$$

The function

$$z = f(x, y) = x^2 + 2xy + 3y^2$$

is homogeneous of degree two. This can be shown by the method above since

$$f(tx, ty) = (tx)^2 + 2(tx)(ty) + 3(ty)^2$$

$$= t^2(x^2 + 2xy + 3y^2) = t^2 z$$

It can also be seen by noticing that if we take each additive term in the expression for z and add the powers of its factors, we find that their sum is 2. By this method it is easy to see that

$$z = f(x, y) = 2 + 2x + 3y$$

is not homogeneous, since the powers of x and y in the first term are 0, and the total power in each of the other terms is 1. Also,

$$f(tx, ty) = 2 + 2tx + 3ty \neq tz$$

A simple example of a homogeneous function of degree one occurs in the production function

$$q = 3L + 5C$$

where q is the level of output resulting from inputs of L units of labour and C units of capital. If both L and C are increased by a factor t the new level of output is

$$q_1 = 3(tL) + 5(tC) = tq$$

so that, for example, doubling the inputs of capital and labour (that is, $t = 2$) doubles the amount of output. This is a case of constant returns to scale.

If the degree of homogeneity k is less than one, there are decreasing returns to scale, and if the degree of homogeneity is greater than one there are increasing returns to scale.

The Cobb–Douglas production function relates output, q, with the amount of labour, L, and capital, C, by the formula

$$q = AL^\alpha C^\beta$$

where A, α, and β are constants. If the inputs of labour and capital are changed by a factor t the new level of output, q_1, is given by

$$q_1 = A(tL)^\alpha (tC)^\beta = At^{\alpha+\beta} L^\alpha C^\beta$$
$$= t^{\alpha+\beta}(AL^\alpha C^\beta) = t^{\alpha+\beta}(q)$$

Hence this function is homogeneous of degree $(\alpha + \beta)$. If $\alpha + \beta = 1$, then $q_1 = tq$ and there are constant returns to scale. For $\alpha + \beta < 1$ there are decreasing returns to scale and for $\alpha + \beta > 1$ there are increasing returns to scale.

7.11 EULER'S THEOREM

For the homogeneous function $z = f(x, y)$ *Euler's theorem* states that

$$xz_x + yz_y \equiv kz$$

where k is the degree of homogeneity and z_x and z_y are the partial derivatives of z.

For example, if

$$z = x^2 + 3y^2 + 2xy$$

then $z_x = 2x + 2y$, $z_y = 6y + 2x$ and $k = 2$

Therefore,
$$xz_x + yz_y = x(2x + 2y) + y(6y + 2x)$$
$$= 2x^2 + 2xy + 6y^2 + 2xy$$
$$= 2(x^2 + 3y^2 + 2xy) = 2z$$

For a homogeneous production function, where z is the level of output and x, y are measures of inputs, z_x and z_y are the marginal products of x and y and so Euler's theorem shows that

(quantity of x) (marginal product of x)

+ (quantity of y) (marginal product of y) = k (total output)

Another application of Euler's theorem occurs in supply theory. If the quantity of a product supplied (q) is dependent upon the price of the product p, and the price of a competing product r, then assuming the function is homogeneous of degree k, Euler's theorem states that

$$p\,\frac{\partial q}{\partial p} + r\,\frac{\partial q}{\partial r} = kq$$

Dividing by q gives

$$\frac{p}{q}\frac{\partial q}{\partial p} + \frac{r}{q}\frac{\partial q}{\partial r} = k$$

that is, the sum of the own price and cross elasticities of supply equal k. For example, if the supply function is

$$q = 6p^2 - 2r^2 - rp$$

$$\frac{\partial q}{\partial p} = 12p - r, \qquad \frac{\partial q}{\partial r} = -4r - p$$

Own price elasticity is

$$\frac{p}{q}\frac{\partial q}{\partial p} = \frac{p}{q}(12p - r)$$

Cross elasticity is

$$\frac{r}{q}\frac{\partial q}{\partial r} = \frac{r}{q}(-4r - p)$$

Sum is

$$\frac{12p^2 - rp}{q} + \frac{(-4r^2) - rp}{q} = \frac{12p^2 - 4r^2 - 2rp}{q} = \frac{2q}{q} = 2$$

which is the degree of homogeneity.

For the Cobb–Douglas function

$$q = AL^{\alpha}C^{\beta}$$

$$q_L = \alpha AL^{\alpha-1}C^{\beta} = \frac{\alpha q}{L}$$

$$q_C = \beta AL^{\alpha}C^{\beta-1} = \frac{\beta q}{C}$$

Hence, by Euler's theorem,

$$Lq_L + Cq_C = \frac{L\alpha q}{L} + \frac{C\beta q}{C} = (\alpha + \beta)q$$

or

$$L\frac{q_L}{\alpha + \beta} + \frac{Cq_C}{\alpha + \beta} = q$$

that is, the total product q is distributed between labour and capital in the proportions

$$\frac{q_L}{\alpha + \beta} : \frac{q_C}{\alpha + \beta} \qquad \text{or} \qquad q_L : q_C$$

and so the distribution is in proportion to the marginal products of labour and capital.

7.12 EXERCISES

1. Show that the following functions are homogeneous and state the degree of homogeneity.

   (a) $f(x, y) = x^3 + 3y^3 - x^2y$

   (b) $f(x, y) = \dfrac{xy + y^2}{4x^3}$

   (c) $f(x, y, z) = 2xyz + x^2y - y^2z$

2. Verify Euler's theorem for the function in 1(a), 1(b) above.
3. The demand for butter depends upon the price of butter (p) and the price of a substitute (r) according to the function

$$q = 10r^2 - 2p^2 + 4rp$$

Show that the sum of the own price and cross elasticities is equal to the degree of homogeneity.

4. Show that the production function $q = 1.1L^{0.6}C^{0.2}$ has decreasing returns to scale and that

$$q = 1.1L^{0.6}C^{0.5}$$

has increasing returns to scale.

(q = output, L = labour, C = capital.)

7.13 MAXIMA AND MINIMA

In the three-dimensional model the function z has a maximum value at the 'top of a hill' and a minimum value at the 'bottom of a valley'. These are the points which are important in economics. A third case exists when z is at a maximum, say, with respect to variation along the x-axis, but is at a minimum with respect to variation along the y-axis. This is called a saddle point since it corresponds to the shape of a saddle: viewed from the side of a horse the saddle is U-shaped, whereas from the front of a horse the saddle is arch-shaped.

To determine the position of these three points it is first necessary to obtain the partial derivative of the function with respect to each variable in turn. A maximum, minimum or saddle point can only exist at a point where each of these partial derivatives is equal to zero. This is a *necessary* condition, eg if $z = f(x, y)$, then for a maximum, minimum or saddle point

$$\frac{\partial z}{\partial x} = 0 \quad \text{and} \quad \frac{\partial z}{\partial y} = 0$$

The solution of these equations simultaneously determines the values of x and y, at which one of these states exists.

To decide whether a particular function has a maximum, a minimum or a saddle point, the second-order partial derivatives must then be considered in the light of the following rules:

1. The function has a maximum or a minimum point if $z_{xx}z_{yy}$ is greater than $(z_{xy})^2$. If z_{xx} and z_{yy} are both negative there is a maximum and if they are both positive there is a minimum.

195

2. If $z_{xx}z_{yy}$ is less than $(z_{xy})^2$ then there is a saddle point if z_{xx} and z_{yy} are both equal to zero.

These rules can be summarised as follows:

Maximum	$z_{xx}z_{yy} > (z_{xy})^2,$	$z_{xx} < 0,$	$z_{yy} < 0$
Minimum	$z_{xx}z_{yy} > (z_{xy})^2,$	$z_{xx} > 0,$	$z_{yy} > 0$
Saddle point	$z_{xx}z_{yy} < (z_{xy})^2,$	$z_{xx} = z_{yy} = 0$	

3. If $z_{xx}z_{yy}$ equals $(z_{xy})^2$ then the point may be a maximum, a minimum, a saddle point or some other type of turning point and requires further investigation.

It can be seen from the above that there are a number of similarities with the conditions for stationary values in the two variable model. Thus, a stationary value can only exist at a point where all the first-order derivatives are equal to zero. For a saddle point the second-order derivatives must also be equal to zero.

If both z_{xx} and z_{yy} are negative (cf $d^2y/dx^2 < 0$) there is a maximum and if both z_{xx} and z_{yy} are positive (cf $d^2y/dx^2 > 0$) there is a minimum.

The procedure for finding maximum and minimum values is, therefore, very simple as the following example shows:

Suppose that $\qquad z = x^2 + 2xy + 2y^2 - 5x - 4y$

then $\qquad z_x = 2x + 2y - 5$

and $\qquad z_y = 2x + 4y - 4$

Equating these partial derivatives to zero produces two simultaneous equations in the variables x and y:

$$2x + 2y = 5$$
$$2x + 4y = 4$$

Subtracting the equations gives

$$2y = -1 \quad \text{or} \quad y = -\tfrac{1}{2} \quad \text{and hence} \quad x = 3$$

There is, therefore, a stationary point when $x = 3$, $y = -\tfrac{1}{2}$.

To determine whether this is a maximum, a minimum or a saddle point, the second-order partial derivatives must be considered:

$$z_{xx} = 2, \qquad z_{yy} = 4, \qquad z_{xy} = 2 = z_{yx}$$

$$z_{xx}z_{yy} = 8 \qquad \text{and} \qquad (z_{xy})^2 = 4$$

$$\therefore \qquad z_{xx}z_{yy} > (z_{xy})^2$$

The point is a maximum or a minimum and as $z_{xx} > 0$, $z_{yy} > 0$ it is a minimum.

The function has a minimum value at the point

$$x = 3, \qquad y = -\tfrac{1}{2}$$

and this value is given by

$$z = 3^2 + (2)(3)(-\tfrac{1}{2}) + 2(-\tfrac{1}{2})^2 - 5(3) - 4(-\tfrac{1}{2})$$

$$= 9 - 3 + \tfrac{1}{2} - 15 + 2 = -6\tfrac{1}{2}$$

7.14 EXERCISES

1. Find z_x, z_y, z_{xx}, z_{xy}, z_{yy} in the following cases:

   (a) $z = x^2 - xy + y^2$ 　　　　　　(b) $z = (x+y)^3$

2. Find the maximum, minimum or saddle-point values (if any) of the following functions:

   (a) $z = y^2 + 2xy + x^3$ 　　　　　　(b) $z = 3x + 2y - 3x^2 - 4y^2$

3. Total cost, z, depends on the output x of product A and the output y of product B.
   What is the minimum value of total cost for the following functions:

   (a) $z = 50 + x^2 + y^2 - 4x - y$

   (b) $z = 200 + x^3 - 3y + y^2 - 2x$

   (c) $z = 250 + x^3 + y^2 - 27x - 8y$

7.15 MAXIMA AND MINIMA FOR FUNCTIONS OF MORE THAN TWO VARIABLES

An extension of this procedure is needed when z is a function of more than two variables because here there are more partial derivatives to deal with.

The first-order conditions for both maxima and minima require all the first partial derivatives of the function to be equal to zero. To distinguish between the maximum and minimum positions we must consider the second-order partial derivatives. This becomes a little more complicated as the number of variables increases and to help make the conditions clear they are stated in determinant notation as explained in Section 1.9.

Consider the case where $z = f(x_1, x_2, x_3, \ldots, x_n)$.

The first-order conditions for a maximum or a minimum are that

$$z_1 = z_2 = z_3 = \cdots = z_n = 0$$

where $$z_1 = \partial z / \partial x_1, \text{ etc}$$

These equations are solved simultaneously to find positions at which the conditions are satisfied. The second-order partial derivatives z_{11}, z_{12}, z_{13}, etc are then calculated and the following determinants formed:

$$\begin{vmatrix} z_{11} \end{vmatrix} \quad \begin{vmatrix} z_{11} & z_{12} \\ z_{12} & z_{22} \end{vmatrix} \quad \begin{vmatrix} z_{11} & z_{12} & z_{13} \\ z_{12} & z_{22} & z_{23} \\ z_{13} & z_{23} & z_{33} \end{vmatrix}$$

The numerical values of these determinants are considered as follows
1. If all the determinants are positive there is a *minimum* position
2. If determinants of even order (ie 2×2, 4×4, etc) are positive but those of odd order (ie 1×1, 3×3, etc) are negative there is a *maximum* position.
3. If any determinants of even order are negative there is a saddle point.

This can be illustrated by an example.

Let $z = f(x_1, x_2, x_3) = 100x_1 - 2x_1^2 + 50x_2 - x_2^2 + 200x_3 - 2x_3^2$

then
$$z_1 = 100 - 4x_1$$
$$z_2 = 50 - 2x_2$$
$$z_3 = 200 - 4x_3$$

If the function is to have a maximum or a minimum then this must occur when
$$z_1 = z_2 = z_3 = 0$$

that is
$$x_1 = 25, \qquad x_2 = 25, \qquad x_3 = 50$$

The second-order partial derivatives are then calculated:

$$z_{11} = -4 \qquad z_{12} = 0 \qquad z_{13} = 0$$
$$z_{22} = -2 \qquad z_{23} = 0 \qquad z_{33} = -4$$

The determinants are

$$\begin{vmatrix} -4 \end{vmatrix} \quad \begin{vmatrix} -4 & 0 \\ 0 & -2 \end{vmatrix} \quad \begin{vmatrix} -4 & 0 & 0 \\ 0 & -2 & 0 \\ 0 & 0 & -4 \end{vmatrix}$$

that is
$$-4, \ +8 \text{ and } -32$$

These values, alternatively negative, positive, negative, satisfy condition 2, and the function therefore has a maximum when the variables have the values $x_1 = 25$, $x_2 = 25$, $x_3 = 50$ and the value of this maximum is given by

$$z = 100(25) - 2(25)^2 + 50(25) - (25)^2 + 200(50) - 2(50)^2 = 6{,}875$$

7.16 CONSTRAINED MAXIMA AND MINIMA

The situation frequently arises when a function has a definite maximum which is not attainable because of some restriction on the value which the variables can assume. This can happen in the case of a production function where it may be impossible to reach the level of output which would yield the maximum revenue due to an insufficient supply of the factors of production. In these cases it is important to determine the maximum which can actually be achieved. This can be done using the method of Lagrangian multipliers.

Let us consider the function

$$z = 100x_1 - 2x_1^2 + 60x_2 - x_2^2$$

To find the maximum value of this without restriction the first-order partial derivatives are obtained and equated to zero.

$$z_1 = 100 - 4x_1 = 0$$

$$z_2 = 60 - 2x_1 = 0$$

$$\therefore \qquad x_1 = 25, \qquad x_2 = 30$$

The second-order partial derivatives are

$$z_{11} = -4 \qquad z_{22} = -2 \qquad z_{12} = 0$$

$$\therefore \qquad z_{11} z_{22} > z_{12}^2$$

and $$\qquad z_{11} < 0 \qquad z_{22} < 0$$

The function, therefore, has a maximum value when $x_1 = 25$, $x_2 = 30$ and $z = 100(25) - 2(25)^2 + 60(30) - (30)^2 = 2,150$.

If there is a limited amount of capital available to buy resources then a constraint is applied to the system. Let this be represented by

$$x_1 + x_2 = 40$$

One method of maximising z subject to this constraint is to substitute directly for x_1, using $x_1 = 40 - x_2$, into z so that

$$z = 100(40 - x_2) - 2(40 - x_2)^2 + 60x_2 - x_2^2$$

$$= 4{,}000 - 100x_2 - 3{,}200 + 160x_2 - 2x_2^2 + 60x_2 - x_2^2$$

$$= 800 + 120x_2 - 3x_2^2$$

This function can be maximised as in Section 5.7 by

$$\frac{dz}{dx_2} = 120 - 6x_2 \qquad \text{and} \qquad \frac{dz}{dx_2} = 0 \qquad \text{when} \qquad x_2 = 20$$

$$\frac{d^2z}{dx_2^2} = -6$$

therefore $x_2 = 20$ gives a maximum and

$$z = 800 + 120(20) - 3(20)^2 = 2,000$$

is the maximum value of z.

In this particular case the method of substituting for x_1 from the constraint into the function for z has worked out well. However, in general such a substitution can be awkward to carry out and also does not produce as much information as an alternative method, using *Lagrangian multipliers*, which is now considered.

This involves forming a new expression.

$$z' = z + \lambda u$$

where $u = (x_1 + x_2 - 40)$ and λ is a constant known as the *Lagrangian multiplier*. It follows because $u = 0$, that by optimising z' we are effectively optimising z and at the same time satisfying the constraint condition. The function is now one in three unknowns and the procedure is to obtain the three first-order partial derivatives and set them equal to zero.

$$z' = 100x_1 - 2x_1^2 + 60x_2 - x_2^2 + \lambda(x_1 + x_2 - 40)$$

$$z_1' = 100 - 4x_1 + \lambda = 0 \tag{1}$$

$$z_2' = 60 - 2x_2 + \lambda = 0 \tag{2}$$

$$z_\lambda' = x_1 + x_2 - 40 = 0 \tag{3}$$

The points at which the function has a maximum or minimum value are obtained by solving this set of simultaneous equations to give $x_1 = 20$, $x_2 = 20$, $\lambda = -20$. To check that this is a maximum value it is necessary to know the values of the second-order partial derivatives of the original function and the first-order partial derivatives of the constraint equation. Again we denote partial derivatives with respect to x_1 by the subscript 1, and partial derivatives with respect to x_2 by the subscript 2. There is a maximum value if

$$2u_1 u_2 z_{12} > u_1^2 z_{22} + u_2^2 z_{11}$$

and the minimum value if

$$2u_1 u_2 z_{12} < u_1^2 z_{22} + u_2^2 z_{11}$$

provided u is linear in both x_1 and x_2.

In the example

$$z_{11} = -4 \qquad z_{12} = 0 \qquad z_{22} = -2$$

$$u_1 = 1 \qquad u_2 = 1$$

Hence $$2u_1 u_2 z_{12} = 0$$

and $$u_1^2 z_{22} + u_2^2 z_{11} = -6$$

and so we have a maximum value when $x_1 = 20$ and $x_2 = 20$. This maximum value is

$$z = 100(20) - 2(20)^2 + 60(20) - (20)^2 = 2,000$$

The maximum output which could be obtained without restriction on x_1 and x_2 was 2,150. Thus, in general, the consequence of placing a constraint upon any of the variables is to reduce the maximum value attainable. An exception would be the case where the two maxima were equal but under these circumstances the constraint is not really effective.

The economic interpretation to be placed on the value of λ is of interest. It is the marginal product of resources, ie it is a measure of the increase in z which would be produced by an increase in the level of the constraint by one unit. For example, if $x_1 + x_2$ can be increased to 41 the function to be maximised is changed to

$$z'' = 100x_1 - 2x_1^2 + 60x_2 - x_2^2 + \lambda(x_1 + x_2 - 41)$$

and in this case it would be found that the maximum value z'' is greater than z' by approximately 20 units (see Section 7.17, Exercise 1).

7.17 EXERCISES
1. Find the maximum value of the function

$$z = 100x_1 - 2x_1^2 + 60x_2 - x_2^2$$

subject to the constraint $x_1 + x_2 = 41$.
2. Find the maximum or minimum value of

$$z = 3x^2 + y^2 - 12x - 14y + 150$$

Find the restricted maximum or minimum value of z when there is a constraint requiring $2x + y = 9$.

3. Find the maximum or minimum value of the following functions subject to the given constraints:

(a) $z = x^2 + y^2$ with $5x + 4y = 40$

(b) $z = 18 - \dfrac{x^2}{2} - y^2$ with $2x + 3y = 12$

(c) $z = 3x^2y^2$ with $x + 3y = 18$

4. Let the production function for a firm be given by

$$x = 20l + 40c - 2l^2 - 3c^2$$

and the cost of labour (l) and of capital (c) be 4 and 5 units respectively.

Find the maximum value of x if the total cost is equal to 28 units.

CHAPTER EIGHT

Linear Programming

8.1 INTRODUCTION

We saw in Chapter 7 how it is possible to find the maximum of a given function when there are constraints on the values which some or all of the variables can assume. To do this we made use of the differential calculus and the method of Lagrangian multipliers. This is, however, only applicable when the constraint is in the form of an equality, that is, one which must be exactly met.

There are many situations where the restriction is not so precise and in particular where the constraint is expressed in the form of an inequality. This might be the case, for example, where a company is concerned with maximising the profitability of its operations, subject to the limitations imposed upon it by the availability of production capacity, whether in the form of machine time, skilled labour or financial resources. This type of optimisation problem cannot generally be handled with the help of the differential calculus, but there are mathematical tools available which can provide a solution in a simple way. To illustrate this we take the case where all the interrelationships between the variables and all the constraints can be expressed in linear form and linear programming can be used. The following example explains the method, and illustrates the principles by means of a graphical approach. This might be used in very simple situations, but these are unlikely to occur in practice and so a mathematical method is explained which has more general applicability and can be computerised easily.

8.2 A PRODUCT-MIX PROBLEM

A small company has two machines X and Y which are both

required to produce two products A and B. The problem is one of deciding how much of each product should be manufactured in order to provide the maximum contribution to the profitability of the company. To obtain a solution it is necessary to have information about the capacity of each machine, the amount of time which is required on each machine by each product, and the contribution which each product makes to profitability. Such a situation might lead to the information provided in Table 8.1.

TABLE 8.1

	Number of hours required per unit of product		Total number of hours available
	A	B	
Machine X	1	2	30
Machine Y	2	1	30
Contribution per unit of product	2	3	

The contribution per unit of product is equal to the selling price less the variable cost of production, and in order to use the linear-programming approach for solving this product-mix problem it is necessary to assume that this remains constant over the levels of output which are under consideration.

If we let x_1 and x_2 be the number of units of products A and B respectively which the company should manufacture, the relevant information can be neatly summarised in algebraic form as follows:

Maximise
$$2x_1 + 3x_2 = z \tag{1}$$

subject to the constraints

$$x_1 + 2x_2 \leqslant 30 \tag{2}$$

$$2x_1 + x_2 \leqslant 30 \tag{3}$$

$$x_1 \geqslant 0 \tag{4}$$

$$x_2 \geqslant 0 \tag{5}$$

205

The first relationship is an equation which represents the total contribution which the company can expect from the two products, ie it is the sum of the contribution per unit multiplied by the number of units which will be manufactured of each product. This is called the *objective function* and the purpose of the exercise is to obtain the values of x_1 and x_2 for which this is a maximum.

The second relationship is an inequality which states that the number of hours used making x_1 units of product A on machine X plus the number of hours used making x_2 units of product B on machine X must not exceed the maximum number of machine hours available. The third relationship is again an inequality and describes the constraint on machine Y in a similar manner. Inequalities (4) and (5) merely state that the output of either product cannot be negative. These are obvious practical constraints on the system.

Any values of the variable x_1 and x_2 which satisfy these inequalities are said to constitute a *feasible solution* to the problem. What is required is a method for determining from all the possible feasible solutions that one which yields the maximum value of z as specified in the objective function. This gives the value of the variables which produce the maximum profit to the company, and is known as the *optimal* solution.

8.3. A GRAPHICAL APPROACH

It is possible to analyse the simple situation under discussion using a graphical method, and this is useful for illustrative purposes, although it is of little use in practical situations where the number of variables is usually much greater.

The approach is to use the two axes of the graph to represent the variables x_1 and x_2. The lines corresponding to the equations $x_1 + 2x_2 = 30$, $2x_1 + x_2 = 30$ are then drawn on the graph as shown in Fig 8.1.

These equations correspond to the upper limit of machine capacity and therefore any combinations of A and B which it is possible to manufacture must lie on the side of the lines closest to the origin. To satisfy both constraints they must, in fact, lie within the area which is doubly cross-hatched in the diagram. This is known as the *feasible region*. We now superimpose on this graph the profit line

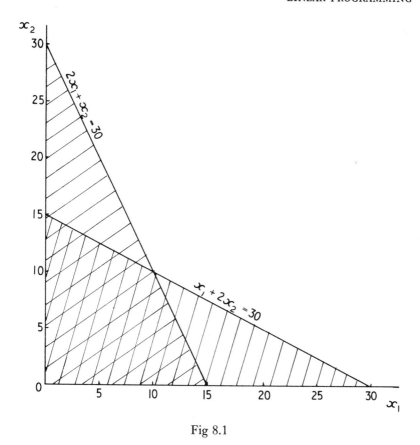

Fig 8.1

$2x_1 + 3x_2 = z$ for profits of 24 and 60 units, that is $2x_1 + 3x_2 = 24$ and 60 respectively.

These lines are parallel to each other because the slope of the line $z = 2x_1 + 3x_2$ is independent of the value of z. They are shown as broken lines in Fig 8.2.

The further away from the origin this line is taken the higher the value of z, but the upper limit which is allowable is determined by the constraints on the system. It is immediately apparent that no point on the line $z = 60$ lies within the feasible region, and therefore this value

207

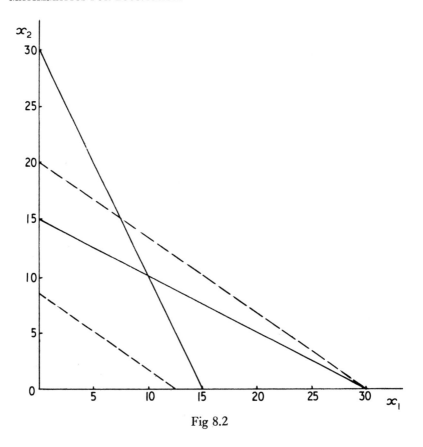

Fig 8.2

of z is unattainable. In fact, the highest value z which satisfies the constraints is reached when the profit line passes through one of the extreme points of the feasible region. This will always be so if the constraints are linear except in the special case where the profit function is parallel to one of the constraints, and in this case, the problem has a multitude of solutions corresponding to the points on the constraint line.

To obtain the solution graphically we move the line with slope $-\frac{2}{3}$ outwards until it reaches the limit of the feasible region, and this point determines the values of x_1 and x_2 corresponding to the outputs of A and B. In this case it is at the point where the two constraint equations intersect, and hence the optimum combination of products to manufacture can be easily determined by the usual method of solution:

$$2x_1 + x_2 = 30 \tag{6}$$

$$x_1 + 2x_2 = 30 \tag{7}$$

Making use of determinants, we have

$$x_1 = \frac{\begin{vmatrix} 30 & 1 \\ 30 & 2 \end{vmatrix}}{\begin{vmatrix} 2 & 1 \\ 1 & 2 \end{vmatrix}} = \frac{60 - 30}{4 - 1} = 10$$

$$x_2 = \frac{\begin{vmatrix} 2 & 30 \\ 1 & 30 \end{vmatrix}}{\begin{vmatrix} 2 & 1 \\ 1 & 2 \end{vmatrix}} = \frac{60 - 30}{4 - 1} = 10$$

The result is that 10 units should be manufactured of both products A and B. It follows that the contribution will equal

$$10 \times 2 + 10 \times 3 = 50$$

It is easy to check that the constraints on machine capacity are not violated by substituting in the original equations

$$x_1 + 2x_2 = 30$$

$$2x_1 + x_2 = 30$$

These equations are exactly satisfied and therefore both machines are fully utilised, ie there is no spare or slack capacity.

8.4 THE SIMPLEX METHOD

The simplex method employs an *iterative procedure* which progresses in a series of steps ultimately leading to the optimal solutions. The rules of operation are described below, using the same example for illustrative purposes.

209

Step 1. Convert the inequalities (2) and (3) into equations by the insertion of new variables x_3 and x_4.

$$x_1 + 2x_2 + x_3 = 30 \tag{8}$$

$$2x_1 + x_2 + x_4 = 30 \tag{9}$$

The only qualification which must be added here is that the new variables must only take on positive values:

$$x_3 \geqslant 0 \qquad x_4 \geqslant 0$$

This is necessary because they represent the difference between the time which is utilised on each machine and the total time available on the machines. They are therefore known as *slack variables* and cannot by definition be negative.

Step 2. Obtain a first feasible solution to the system. The insertion of the slack variables proves useful here because an immediately obvious solution is given by

$$x_1 = 0, \qquad x_2 = 0, \qquad x_3 = 30, \qquad x_4 = 30$$

A unique solution to two equations in four unknowns can only be obtained if two of the unknowns are given defined values. For convenience they are assigned the value zero in this exercise.

This solution corresponds to zero output and slack or spare capacity on both machines and also produces zero profit. It corresponds to the origin in the graphical method and hence it is on the edge of the feasible region. The reason it is chosen is that it is an easily found solution. The non-zero variables x_3 and x_4 are taken as a starting point for the iterative procedure and are known as the *basic elements* and form a *basis* in a tableau which is drawn up in Table 8.2.

TABLE 8.2

	x_1	x_2	x_3	x_4	p
x_3	1	2	1	0	30
x_4	2	1	0	1	30
z	-2	-3	0	0	0

The elements in the tableau are obtained from the initial data for the system. They correspond to the coefficients of equations (8) and (9) and of the objective function equated to zero.

Equation (8) $$x_1 + 2x_2 + x_3 + 0x_4 = 30$$

Equation (9) $$2x_1 + x_2 + 0x_3 + x_4 = 30$$

Objective function $$z - 2x_1 - 3x_2 + 0x_3 + 0x_4 = 0$$

In the tableau the basic elements x_3 and x_4 are written to the left of the equation in which they appear.

The first feasible solution is obtained by letting $x_1 = 0$ and $x_2 = 0$ with the result that .

$$x_3 = 30 \qquad \text{[from eq. (8)]}$$

$$x_4 = 30 \qquad \text{[from eq. (9)]}$$

and the profit, $\qquad z_1 = 0 \qquad$ (from the objective function)

It is obvious that this solution can be improved upon and that any improvements can only come from the manufacture of either A or B. This could be shown in the tableau by either x_1 or x_2 appearing in the basis with the result that either x_3 or x_4 must become zero and leave the basis.

The exercise consists of:

(a) determining the variable to enter the basis, known as the *entering* variable, and,

(b) determining the variable to leave the basis, known as the *departing* variable, and

(c) calculating the profit which results from the interchange.

One such change is made at each state in the iterative procedure and if the rules are followed carefully the method leads progressively to the optimum solution and indicates clearly when this optimum is reached.

Step 3. Select the entering variable by considering the elements in the lowest row of the tableau. These elements correspond to the marginal profit which would be obtained by the manufacture of any products other than those at present in the basis. In the first tableau the elements are -2 and -3, which correspond to the contribution per unit from products A and B respectively. It is possible to introduce only one new variable into the basis at a time and it seems reasonable to select that variable which offers the largest marginal profit, in this case x_2, or product B. Thus we have

Rule 1. The entering variable is chosen as that variable which has the largest negative element in the bottom row of the tableau.

Step 4. Select the departing variable. Before doing this, it is important to remember that the variable leaving the basis becomes zero and the two *non-basic* zero variables determine the values of the basic variables. For example,

if $$x_3 = 0 \quad \text{and} \quad x_1 = 0$$

then $$2x_2 = 30 \qquad \text{[from eq. (8)]}$$

and $$x_2 + x_4 = 30 \qquad \text{[from eq. (9)]}$$

That is $$x_2 = 15 \qquad x_4 = 15$$

But if $$x_4 = 0 \quad \text{and} \quad x_1 = 0$$

then $$2x_2 + x_3 = 30 \qquad \text{[from eq. (8)]}$$

$$x_2 = 30 \qquad \text{[from eq. (9)]}$$

That is $$x_2 = 30 \qquad x_3 = -30$$

This second result is not feasible because the values of the variables in a linear programming problem must all be positive and $x_3 = -30$ violates this constraint.

It follows therefore that the departing variable can only be x_3.

In the case of large problems, it would be a laborious task to check each variable independently to see whether the resulting interchange violates the non-negativity constraints. It is not, in fact, necessary because the departing variable can be found by considering the ratio p_i/a_{ij} for each basic variable. p_i is the element in the last column of the row in which the basic variable is written and a_{ij} is the element in the column of the entering variable and the row of the basic variable.

The calculation for the present example is as follows:

	x_1	x_2	x_3	x_4	p	$\dfrac{p_i}{a_{ij}}$
x_3	1	2	1	0	30	$\dfrac{30}{2}$
x_4	2	1	0	1	30	$\dfrac{30}{1}$
z	-2	-3	0	0	0	

$\uparrow$

entering variable

In order to ensure that all the variables will have positive values at each state of the procedure, it is necessary to specify

Rule 2. The departing variable is that one for which the ratio p_i/a_{ij} has the *smallest positive* value.

The element at the intersection of the column containing the entering variable and the row containing the departing variable is known as the *pivot* and plays an important part in the following step.

In this example the entering variable is x_2, the departing variable is x_3 and the pivot element is 2.

	x_1	x_2	x_3	x_4	p	
x_3	1	2	1	0	30	←departing variable
x_4	2	1	0	1	30	
z	-2	-3	0	0	0	

$\uparrow$

entering variable

We now require to form the new set of equations from which we can determine the value of x_2 when x_1 and x_3 have the value zero. This will tell us how many units of product A to manufacture and how many units of spare capacity there will be. This information, along with the profit which will be produced by this course of action, can be obtained in the following way.

Step 5. Rewrite the tableau with the entering variable replacing the departing variable in the basis and form the new matrix of coefficients as follows:

213

(a) Divide all the elements in the pivot row by the pivot element.

(b) Add or subtract such multiples of the element in the pivot row from corresponding elements of the other rows to reduce all other elements in the pivot column to zero.

In this example all the elements in the pivot row are divided by two to form a new row. Multiples of this new row are added or subtracted from the other rows as follows:

(1) The elements of the new row are subtracted from the corresponding elements of row two of the old tableau to form row two of the new tableau.

(2) Three times the elements of the new row are added to the corresponding elements of row three of the old tableau to form row three of the new tableau.

	x_1	x_2	x_3	x_4	p
x_2	$\frac{1}{2}$	1	$\frac{1}{2}$	0	15
x_4	$(2-\frac{1}{2})$	$(1-1)$	$(0-\frac{1}{2})$	$(1-0)$	$(30-15)$
z	$(-2+\frac{3}{2})$	$(-3+3)$	$(0+\frac{3}{2})$	$(0+0)$	$(0+45)$

This can be rewritten as

	x_1	x_2	x_3	x_4	p
x_2	$\frac{1}{2}$	1	$\frac{1}{2}$	0	15
x_4	$\frac{3}{2}$	0	$-\frac{1}{2}$	1	15
z	$-\frac{1}{2}$	0	$\frac{3}{2}$	0	45

This tableau corresponds to the solution

$$x_1=0, \quad x_2=15, \quad x_3=0, \quad x_4=15, \quad z=45$$

that is, the company could manufacture 15 units of B and as each unit of B contributes 3 units to the profit and overhead, the total contribution would amount to 45 units.

It can also be noted that this product mix uses all the available capacity of machine X (slack variable $x_3=0$) but leaves spare capacity on machine Y (slack variable $x_4 = 15$). Check: 15 units of product B require $15 \times 2 = 30$ hours on machine X but only $15 \times 1 = 15$ hours on machine Y.

The second tableau gives a solution which is an improvement on that given by the first tableau. But is it the optimal solution? The answer is no, and this is indicated by the presence in the bottom row of the tableau of a negative element. These elements correspond to the marginal profit which could be obtained by the introduction into the basis of a non-basic element. Therefore, in this example, profit is being foregone by not having x_1 in the basis, ie by not manufacturing product A. This profit is equal to $\frac{1}{2}$ unit for each unit of x_1 introduced. It should be noted here that this marginal profit is no longer equal to the direct contribution per unit from A because the introduction of this product necessarily means that we must reduce the quantity of product B which is produced in order not to violate the capacity constraint on machine X. (There was no spare capacity on this machine with the first product mix.) The negative figures in the bottom row of the tableau therefore correspond to the marginal profit which is obtained after the product mix has been changed by introducing product A and adjusting the level of manufacture of the other product (or products) to satisfy the constraints on the system.

The variable with the *largest negative element* in the bottom row of the tableau is selected for the entering variable (in this case x_1 which has the only negative element). The departing variable is found by considering the ratio p_i/a_{ij} and the interchange carried out as before.

Steps 3 and 4 Repeated

	x_1	x_2	x_3	x_4	p	$\dfrac{p_i}{a_{ij}}$
x_2	$\frac{1}{2}$	1	$\frac{1}{2}$	0	15	$15/\frac{1}{2}$
x_4	$\boxed{\frac{3}{2}}$	0	$-\frac{1}{2}$	1	15	$15/\frac{3}{2}$ ←departing variable
z	$-\frac{1}{2}$	0	$\frac{3}{2}$	0	45	

$\uparrow$
entering variable

215

Step 5 Repeated

	x_1	x_2	x_3	x_4	p
x_2	$(\frac{1}{2}-\frac{1}{2})$	$(1-0)$	$(\frac{1}{2}+\frac{1}{6})$	$(0-\frac{1}{3})$	$(15-5)$
x_1	1	0	$-\frac{1}{3}$	$\frac{2}{3}$	10
z	$(-\frac{1}{2}+\frac{1}{2})$	$(0+0)$	$(\frac{3}{2}-\frac{1}{6})$	$(0+\frac{1}{3})$	$(45+5)$

which is rewritten as

	x_1	x_2	x_3	x_4	p
x_2	0	1	$\frac{2}{3}$	$-\frac{1}{3}$	10
x_1	1	0	$-\frac{1}{3}$	$\frac{2}{3}$	10
z	0	0	$\frac{4}{3}$	$\frac{1}{3}$	50

This tableau corresponds to the solution

$$x_1 = 10, \quad x_2 = 10, \quad x_3 = 0, \quad x_4 = 0, \quad z = 50$$

that is, the company would manufacture 10 units of A, 10 units of B, and the total contribution from this combination of output would be 50. The last iteration has again improved the profit of the company and at the same time has arrived at a combination which utilises all the available capacity of both the scarce resources (slack variables x_3 and x_4 are both equal to zero). It can also be seen that none of the elements in the bottom row of the tableau are negative and this indicates that no increase in profit can be obtained by a change in the product mix. The final tableau therefore provides the optimal solution to the problem.

8.5 A SUMMARY OF THE SIMPLEX METHOD

To explain and carry out the iterative procedure at one and the same time may give the impression that it is lengthy and difficult. To show that this is not so, the necessary stages are repeated here with all the detail that would be required to carry out the calculations in a practical case. The problem itself has been altered slightly to one in which the company has more capacity available on machine X than on machine Y. The data for the problem is now as shown in Table 8.3.

TABLE 8.3

	Number of hours required per unit of product		Total number of hours available
	A	B	
Machine X	1	2	40
Machine Y	2	1	30
Contribution per unit of product	2	3	

Mathematically this can be stated as follows.

Maximise $\qquad\qquad 2x_1 + 3x_2 = z$

subject to the constraints

$$x_1 + 2x_2 \leqslant 40$$
$$2x_1 + x_2 \leqslant 30$$
$$x_1 \geqslant 0$$
$$x_2 \geqslant 0$$

The slack variables, x_3 and x_4 are inserted in the inequalities with the qualification that $x_3 \geqslant 0$, $x_4 \geqslant 0$

$$x_1 + 2x_2 + x_3 = 40$$
$$2x_1 + x_2 + x_4 = 30$$

The procedure is as shown in Tableaux 1–3. The optimal solution is to make $6\frac{2}{3}$ units of A and $16\frac{2}{3}$ units of B; the total contribution from this combination would be $63\frac{1}{3}$. This solution assumes that it is possible to manufacture fractional parts of a product and in many cases this is not unrealistic because it is the average output per week which is being discussed, and this is very likely to include fractional units. When such a situation is not permissible it is possible to continue the procedure further and use the method of integer programming which will produce a solution in which the variables are restricted to whole units.

217

Tableau 1

	x_1	x_2	x_3	x_4	p	$\dfrac{p_i}{a_{ij}}$	
x_3	1	$\boxed{2}$	1	0	40	40/2	←departing variable
x_4	2	1	0	1	30	30/1	
z	-2	-3	0	0	0		
		↑					

entering variable

Tableau 2

	x_1	x_2	x_3	x_4	p	$\dfrac{p_i}{a_{ij}}$	
x_2	$\frac{1}{2}$	1	$\frac{1}{2}$	0	20	$20/\frac{1}{2}$	
x_4	$\boxed{\frac{3}{2}}$	0	$-\frac{1}{2}$	1	10	$10/\frac{3}{2}$	←departing variable
z	$-\frac{1}{2}$	0	$\frac{3}{2}$	0	60		
	↑						

entering variable

Tableau 3

	x_1	x_2	x_3	x_4	p
x_2	0	1	$\frac{2}{3}$	$-\frac{1}{3}$	$\frac{50}{3}$
x_1	1	0	$-\frac{1}{3}$	$\frac{2}{3}$	$\frac{20}{3}$
z	0	0	$\frac{4}{3}$	$\frac{1}{3}$	$\frac{190}{3}$

218

8.6 INCREMENTAL VALUES

At a very early stage in the problem under discussion, it was assumed, without being stated, that the available capacity of the scarce resources, ie machines X and Y, would not be increased. This need not always be so, and an increase might be possible in a variety of ways, of which the investment in new machines or the addition of overtime working to the present machines are two obvious approaches to the problem. By conveniently ignoring this aspect of the problem, it was possible to work throughout entirely in terms of variable costs and to ignore the fixed-cost element, which is only of importance when increases or reductions in capacity are contemplated. When such changes are being considered it is important to know where and in what quantity the investment in new equipment should be made.

It would be inappropriate at this point not to mention and make use of some further information which is obtained as a by-product of the linear programme. This information appears in the bottom row of the final tableau and is the *incremental worth* of the scarce resources or more simply the extra profit which could be earned by increasing the availability of any scarce resource by one unit.

Tableau 4

	x_1	x_2	x_3	x_4	p
x_2	0	1	$\frac{2}{3}$	$-\frac{1}{3}$	10
x_1	1	0	$-\frac{1}{3}$	$\frac{2}{3}$	10
z	0	0	$\frac{4}{3}$	$\frac{1}{3}$	50

Returning to the first example, the final tableau is as shown in Tableau 4. The bottom row contains zero elements in the columns corresponding to the variables in the basis, that is, x_1 and x_2, but positive elements in the columns corresponding to the non-basic variables, that is, x_3 and x_4. These positive elements correspond to the incremental value of one hour of machine time on X and Y respectively. This means that one extra hour of available machine time on X would increase the total contribution by $\frac{4}{3}$ to $51\frac{1}{3}$ and one extra machine hour on Y would increase it by $\frac{1}{3}$ to $50\frac{1}{3}$.

In many cases, of course, extra capacity cannot be added in single units, eg hours, but must be added in fairly large steps. This would occur where the purchase of an extra machine was contemplated with a resulting increase of capacity of, say, 30 hours. Occasionally the structure of the problem allows the increased contribution to be determined directly from the incremental value. In the second example, an increase in capacity of 10 hours in machine X resulted in an increased contribution of $13\frac{1}{3}$ units, and this is exactly equal to $10 \times \frac{4}{3}$, ie the extra contribution was directly proportional to the number of additional units of capacity. This is not generally the case, and therefore it is better to re-run the linear programme if the effect of changes in any of the constraints is required. This technique, included under the general name of *parametric programming*, is used by many industries at the present time.

A final interesting point which arises from the solution of the linear programme is, as economists might expect, that at the optimum the marginal revenue equals the marginal cost for all those products which the company should produce.

In the example, the incremental value of machine capacity of X and Y is respectively $\frac{4}{3}$ and $\frac{1}{3}$, and this must therefore be the value to the company of time on these machines at the margin. By multiplying these values by the hours which a unit of A and B uses on the machines we will obtain a value for the marginal cost of the two products.

Product A: marginal cost $= 1 \times \frac{4}{3} + 2 \times \frac{1}{3} = \frac{6}{3} = 2$

The contribution per unit of A which is in fact its marginal revenue is also equal to 2.

Product B: marginal cost $= 2 \times \frac{4}{3} + 1 \times \frac{1}{3} = \frac{9}{3} = 3$

This is equal to the contribution per unit or marginal revenue from B. It can be quite easily shown that the marginal revenue equals the marginal cost for all products that appear at a positive level in the optimal solution. Those products which do not appear in the final solution have a marginal cost greater than their marginal revenue, which is a result that would be expected on theoretical grounds.

8.7 THE GENERAL PROBLEM

It is not difficult to extend the method of analysis to the case where three products can be manufactured and the objective function can be expressed as a linear combination of the contribution which each of these products makes to profit and overhead.

For example, maximise

$$z = f(x_1, x_2, x_3) = p_1 x_1 + p_2 x_2 + p_3 x_3$$

subject to the constraints of the form

$$a_1 x_1 + a_2 x_2 + a_3 x_3 \leqslant M \qquad \text{and} \qquad x_i \geqslant 0$$

A further extension can be made to the general case:

Maximise $$z = f(x_1, x_2, \ldots, x_n)$$

subject to the constraints

$$\sum_{i=1}^{n} a_{ij} x_i \leqslant M_j \qquad \text{and} \qquad x_i \geqslant 0$$

where $$\sum_{i=1}^{n} a_{ij} x_i = a_{1j} x_1 + a_{2j} x_2 + \cdots + a_{nj} x_n$$

$j = 1$ to m, and there are m constraints in the system.

The general case can be neatly expressed using matrix notation as was shown in Chapter 2.

8.8 EXERCISES

1. Use the simplex method to establish the product mix which will provide the maximum contribution in the example shown in Table 8.4. Calculate the additional contribution which will be obtained if one additional hour of capacity became available on either machine X or machine Y and show that this is equal to the incremental value of machine capacity derived from the final tableau of the simplex procedure.

2. Using the basic data provided in Exercise 1, determine the number of units of A and B which should be manufactured if either of the following conditions applies.

221

TABLE 8.4

| | Number of hours required per unit of product | | Total number of hours available |
	A	B	
Machine X	2	3	25
Machine Y	4	1	35
Contribution per unit of product	9	7	

(a) the total number of available hours of machine Y is increased to (i) 40 and (ii) 45 in successive stages;
(b) the contribution per unit of A is reduced to 4;
(c) the number of hours required by product B on machine X is reduced to 2.

CHAPTER NINE

Differential Equations

9.1 INTRODUCTION
In Chapter 6 we saw that equations of the form

$$\frac{dy}{dx} = f(x)$$

could be solved by integration. Such an equation is known as a *differential equation* since it expresses the relationship between the derivative of y with respect to x and the value of x itself. It contains only the first-order derivative and is, therefore, known as a *first-order differential equation* and because this term is only raised to the power one it is a *first-degree* equation.

$(dy/dx)^2 = 6x + 2$ is a differential equation of *first-order and second-degree* because it contains only a first-order derivative and this is raised to the power two.

$d^2y/dx^2 = 6$ is a differential equation of *second-order and first-degree* because it contains a second-order derivative which is raised to the power one.

$$\frac{d^2y}{dx^2} + \left(\frac{dy}{dx}\right)^2 + y = 0 \qquad \text{and} \qquad \frac{d^2y}{dx^2} - y = 0$$

are both differential equations of second-order and first-degree because the highest-order derivative that is present in each equation is the second and this is raised to the power one in both cases.

In general a differential equation contains one or more derivatives all or some of which may be raised to a power other than zero.

Differential equations are classified into groups by

1. *The order*. This is the same as that of the highest-order derivative which is present in the equation.

2. *The degree*. This is determined by the power to which the highest-order derivative is raised.

The general differential equation

$$\left(\frac{d^n y}{dx^n}\right)^\kappa + \left(\frac{d^{n-1} y}{dx^{n-1}}\right)^\theta + \cdots + y + c = 0$$

is one of the nth order and the κth degree, even if θ is greater than κ.

The equations which occur most frequently are those of first-and second-order, both usually being of the first degree. The problem is to find a solution which satisfies the following conditions.

1. It must be free of terms containing derivatives.

2. It must satisfy the differential equation.

This can be illustrated by reference to the example

$$\frac{d^2 y}{dx^2} = 6$$

Integration of this second-order equation leads to the first-order equation

$$\frac{dy}{dx} = 6x + A$$

which on further integration leads to

$$y = 3x^2 + Ax + B$$

The final equation expresses y as a function of x and it has no terms containing derivatives. It also satisfies the original second-order differential equation, as can be easily checked by differentiation. It satisfies both of the necessary conditions stated above and is, therefore, a solution to the differential equation. The function was obtained by integration, in this case in two stages, and the procedure which is adopted generally for finding the solution to a differential equation is one of integration.

Differential equations can and do arise in a wide variety of situations and for this reason they appear in many different forms. In order to develop and use a standard set of techniques for obtaining a solution to such equations they are grouped together under a number of headings. The method of integrating these functions is discussed here with particular reference to the methods which are available for finding a solution to the more common type of problem.

9.2 FIRST-ORDER LINEAR DIFFERENTIAL EQUATIONS
The most general form of first-order linear equation is

$$\frac{dy}{dx} + P(x)y = Q(x)$$

where $P(x)$, $Q(x)$ are either functions of x only, or constants.

We will deal with various special cases before looking at a general method of solution of this equation.

9.2.1. VARIABLES SEPARABLE
An equation with separable variables is of the form

$$\frac{dy}{dx} = f(x, y)$$

where the functions can be rewritten in the form

$$f_1(x) = f_2(y)\frac{dy}{dx}$$

with $f_1(x)$ containing only terms in x and $f_2(y)$ containing only terms in y.

For example, if

$$\frac{dy}{dx} = \frac{x}{y}$$

then it is possible to cross-multiply with the result

$$y\,dy = x\,dx$$

Integrating with respect to x, we obtain

$$\int f_1(x)\,dx = \int f_2(y)\frac{dy}{dx}\,dx = \int f_2(y)\,dy$$

225

The solution can be found by integrating both sides of this equation

$$\int y \, dy = \int x \, dx$$

$$\therefore \qquad \tfrac{1}{2}y^2 = \tfrac{1}{2}x^2 + C$$

or $\qquad y^2 - x^2 = 2C = \text{constant} = A$

A check can be made that this solution does satisfy the differential equation by differentiating, in the following manner. (For discussion see Section 7.8.)

Let $\qquad z = y^2 - x^2 - A$

then $\qquad \dfrac{\partial z}{\partial x} = -2x \qquad \dfrac{\partial z}{\partial y} = 2y$

and $\qquad \dfrac{dy}{dx} = -\dfrac{\partial z/\partial x}{\partial z/\partial y} = -\dfrac{(-2x)}{2y} = \dfrac{x}{y}$

the solution does therefore, satisfy the equation. The arbitrary constant, A, arises because the process is simply one of integration. To determine the value of this constant requires some further information. For example, if

$$y = 2 \qquad \text{when} \qquad x = 1$$

then $\qquad 2^2 - 1^2 = A \qquad \text{or} \qquad A = 3$

and the unique solution to the differential equation

$$\dfrac{dy}{dx} = \dfrac{x}{y} \qquad \text{given} \qquad y = 2 \quad \text{when} \quad x = 1$$

is $\qquad y^2 - x^2 = 3$

Example What is the general form of the demand equation which has a constant elasticity of -1?

Let q be the quantity demanded at price p. Then elasticity, e, is defined by

$$e = \frac{p}{q} \frac{dq}{dp}$$

For $e = -1$, we require

$$\frac{p}{q}\frac{dq}{dp} = -1 \quad \text{or} \quad \frac{dq}{q} = -\frac{dp}{p}$$

Integrating, we have

$$\log_e (q) = -\log_e (p) + \log_e (A)$$

where A is a constant.

$$\therefore \qquad \log_e (q) = \log_e (p^{-1}) + \log_e (A) = \log_e (Ap^{-1})$$

and $\qquad\qquad q = Ap^{-1} \quad \text{or} \quad pq = A$

9.2.2. LINEAR DIFFERENTIAL EQUATIONS WITH HOMOGENEOUS COEFFICIENTS

The above method breaks down if the variables are not separable but if the differential equation

$$f_1(xy)\, dx + f_2(xy)\, dy = 0$$

has homogeneous coefficients it can be converted to the variables-separable type by a suitable change of variable. (For a discussion of homogeneity see Section 7.10.)

For example, if

$$\frac{dy}{dx} = \frac{x+y}{x}$$

then $\qquad\qquad (x+y)\, dx - x\, dy = 0$

In this case the variables are not separable but the coefficients of dx and dy are homogeneous of degree one. This can be converted into the variables-separable type by the substitutions

$$y = vx \quad \text{and} \quad dy = v\, dx + x\, dv$$

After carrying out the substitution we have

$$(x+vx)\, dx - x(v\, dx + x\, dv) = 0$$

which can be rearranged into

$$x(1-v)\, dx + x(v\, dx - x\, dv) = 0$$

$$x[(1-v)\, dx + (v\, dx - x\, dv)] = 0$$

227

If this is to be valid for all values of x then

$$(1 - v)\, dx + (v\, dx - x\, dv) = 0$$

that is

$$dx - v\, dx + v\, dx - x\, dv = 0$$

or

$$dx - x\, dv = 0$$

This is now in the variables-separable form and can be treated as such

$$dx = x\, dv$$

$$\frac{dx}{x} = dv$$

$$\therefore \qquad \int \frac{dx}{x} = \int dv$$

and hence

$$\log_e x = v + c$$

To obtain the solution in terms of the variables x and y it is necessary to substitute for v in this equation.

$$v = \frac{y}{x}$$

$$\therefore \qquad \log_e x = \frac{y}{x} + c \qquad \text{or} \qquad y = x(\log_e x - c)$$

This is, therefore, the solution of the linear differential equation

$$\frac{dy}{dx} = \frac{x + y}{x}$$

The result can be checked by differentiation.

Example The total cost, y, of producing x units of output of a product is related to the marginal cost of production by the equation

$$\frac{dy}{dx} = \frac{y - 2x}{2y - x}$$

What is the total cost function?

The equation can be re-written as

$$(2y - x)\, dy + (2x - y)\, dx = 0$$

The coefficients of this equation are homogeneous of degree 1, and so we substitute $y = vx$ and $dy = v\,dx + x\,dv$

$$(2vx - x)(v\,dx + x\,dv) + (2x - vx)\,dx = 0$$

$$(2v^2x - xv + 2x - vx)\,dx + (2vx^2 - x^2)\,dv = 0$$

Dividing through by x gives

$$(2v^2 - 2v + 2)\,dx + x(2v - 1)\,dv = 0$$

that is

$$\frac{dx}{x} + \frac{2v - 1}{2v^2 - 2v + 2}\,dv = 0$$

Integrating gives

$$\log_e x + \tfrac{1}{2}\log_e(2v^2 - 2v + 2) = \log_e A$$

or

$$x(2v^2 - 2v + 2)^{1/2} = A$$

Substituting $v = y/x$ and squaring gives

$$x^2\left(\frac{2y^2}{x^2} - \frac{2y}{x} + 2\right) = A^2$$

or

$$2y^2 - 2xy + 2x^2 = A^2$$

This is the relationship between total cost, y, and output x.

9.2.3. LINEAR DIFFERENTIAL EQUATIONS WITH NON-HOMOGENEOUS COEFFICIENTS

These are of the form

$$(a_1x + b_1y + c_1)\,dy + (a_2x + b_2y + c_2)\,dx = 0$$

The coefficients are not homogeneous but they can be made so by the substitutions

$$a_1x + b_1y + c_1 = Y \tag{1}$$

$$a_2x + b_2y + c_2 = X \tag{2}$$

This results in

$$X\,dx + Y\,dy = 0$$

or

$$\frac{dy}{dx} = -\frac{X}{Y} \tag{3}$$

It is now necessary to differentiate eqs. (1) and (2)

$$\frac{dY}{dx} = a_1 + b_1 \frac{dy}{dx} = a_1 + b_1 \left(-\frac{X}{Y} \right) \tag{4}$$

$$\frac{dX}{dx} = a_2 + b_2 \frac{dy}{dx} = a_2 + b_2 \left(-\frac{X}{Y} \right) \tag{5}$$

Dividing eq. (4) by eq. (5), we obtain

$$\frac{dY/dx}{dX/dx} = \frac{dY}{dX} = \frac{a_1 + b_1(-X/Y)}{a_2 + b_2(-X/Y)}$$

$$\therefore \qquad \frac{dY}{dX} = \frac{a_1 Y - b_1 X}{a_2 Y - b_2 X}$$

or $\qquad (a_1 Y - b_1 X)\, dX - (a_2 Y - b_2 X)\, dY = 0$

This equation is homogeneous in X and Y and it is possible to obtain a solution by the method of Section 9.2.2. The exercise is left to the reader to complete as the procedure is no different from that previously demonstrated (Section 9.3, Exercise 1).

Example $\qquad (x + 2y + 1)\, dx + (2x + y + 2)\, dy = 0$

Let $\qquad\qquad\qquad x + 2y + 1 = X \tag{1 a}$

$$2x + y + 2 = Y \tag{2 a}$$

and $\qquad\qquad\qquad \frac{dy}{dx} = -\frac{X}{Y} \tag{3 a}$

Differentiating eqs. (1 a) and (2 a), we have

$$\frac{dX}{dx} = 1 + \frac{2dy}{dx} = 1 - \frac{2X}{Y} = \frac{Y - 2X}{Y} \tag{4 a}$$

$$\frac{dY}{dx} = 2 + \frac{dy}{dx} = 2 - \frac{X}{Y} = \frac{2Y - X}{Y} \tag{5 a}$$

Dividing eq. (5 a) by eq. (4 a) gives

$$\frac{dY/dx}{dX/dx} = \frac{dY}{dX} = \frac{(2Y - X)/Y}{(Y - 2X)/Y} = \frac{2Y - X}{Y - 2X}$$

$$\therefore \qquad (2Y - X)\, dX - (Y - 2X)\, dY = 0$$

This is homogeneous in X and Y, so let

$$Y = VX \quad \text{and} \quad dY = V \, dX + X \, dV$$

then

$$(2VX - X) \, dX - (VX - 2X)(V \, dX + X \, dV) = 0$$

$$X[(2V - 1) \, dX - (V - 2)(V \, dX + X \, dV)] = 0$$

$$\therefore \quad (2V - 1) \, dX - (V - 2)(V \, dX + X \, dV) = 0$$

$$2V \, dX - dX - V^2 \, dX - VX \, dV + 2V \, dX + 2X \, dV = 0$$

$$(2V - 1 - V^2 + 2V) \, dX - (VX - 2X) \, dV = 0$$

$$(-V^2 + 4V - 1) \, dX - X(V - 2) \, dV = 0$$

This is in the variables-separable form and can be written

$$\frac{dX}{X} = \left(\frac{V - 2}{-V^2 + 4V - 1} \right) dV$$

The integral of the left-hand side is simply $\log_e X$. The integral of the right-hand side is obtained as follows:

$$\int \frac{V - 2}{-V^2 + 4V - 1} \, dV = -\int \frac{\frac{1}{2}(2V - 4)}{V^2 - 4V + 1} \, dV$$

$$= -\frac{1}{2} \int \frac{2V - 4}{V^2 - 4V + 1} \, dV$$

$$= -\frac{1}{2} \log_e (V^2 - 4V + 1) + C$$

Combining these two results

$$\log_e X = -\frac{1}{2} \log_e (V^2 - 4V + 1) + C$$

It is more convenient in examples such as this to let the constant assume a logarithmic form, ie let $C = \log_e A$. From the properties of logarithms (see Section 4.14) it is possible to write this as

$$\log_e X = \log_e A - \log_e \sqrt{(V^2 - 4V + 1)}$$

$$= \log_e \frac{A}{\sqrt{(V^2 - 4V + 1)}}$$

$$\therefore \quad X = \frac{A}{\sqrt{(V^2 - 4V + 1)}}$$

231

or $$X^2(V^2 - 4V + 1) = A^2 = \text{constant} = B$$

The result must be expressed in terms of the variables x and y and this is done in two stages

1. Replace V by Y/X

then $$X^2 \left(\frac{Y^2}{X^2} - \frac{4Y}{X} + 1 \right) = B$$

that is $Y^2 - 4YX + X^2 = B$

2. Replace X by $(x + 2y + 1)$ and Y by $(2x + y + 2)$

then $(2x + y + 2)^2 - 4(x + 2y + 1)(2x + y + 2) + (x + 2y + 1)^2 = B$

that is,

$$4x^2 + y^2 + 4 + 4xy + 4y + 8x - 8x^2 - 4xy - 8x - 16xy - 8y^2 - 16y$$
$$- 8x - 4y - 8 + x^2 + 4y^2 + 1 + 4xy + 4y + 2x = B$$

or $$- 3x^2 - 3y^2 - 3 - 12xy - 12y - 6x = B$$

$$- 3(x^2 + y^2 + 4xy + 4y + 2x + 1) = B$$

$\therefore$ $$x^2 + y^2 + 4xy + 4y + 2x = -\frac{B}{3} - 1 = \text{constant}$$

$\therefore$ the solution to the differential equation

$$(x + 2y + 1)\,dx + (2x + y + 2)\,dy = 0$$

is given by

$$x^2 + y^2 + 4xy + 4y + 2x = C$$

9.2.4. EXACT DIFFERENTIAL EQUATIONS

An *exact differential equation* involving x and y is one which may immediately (ie without multiplying through by any factor) be expressed in terms of derivatives of functions of x and y, where these functions do not themselves involve derivatives.

$$3x^2y + x^3 \frac{dy}{dx} = 0$$

is an exact differential equation because it is formed by differentiating the equation $x^3y = A$.

It should be possible to find the solution to an exact differential equation very quickly if it is recognised as such. This is not always obvious at first sight but the following rule always holds.

$$P \, dx + Q \, dy = 0$$

is an exact differential equation if

$$\frac{\partial P}{\partial y} = \frac{\partial Q}{\partial x}$$

For example, consider the equation

$$3x^2 y + x^3 \frac{dy}{dx} = 0$$

$$(3x^2 y) \, dx + (x^3) \, dy = 0$$

$$\therefore \qquad \frac{\partial(3x^2 y)}{\partial y} = 3x^2 \quad \text{and} \quad \frac{\partial(x^3)}{\partial x} = 3x^2$$

These two partial derivatives are equal and therefore the equation is exact. The solution in this case is fairly obvious when the values of P and Q are considered.

$$P = 3x^2 y \qquad Q = x^3$$

and the solution is $x^3 y = C$.

When the solution is not so obvious it can be found by using the following relationships:

If
$$z = f(x, y) = 0$$

Then
$$dz = \frac{\partial z}{\partial x} \, dx + \frac{\partial z}{\partial y} \, dy = 0$$

(See Section 7.5.)

To reverse this procedure and find z it is necessary to compare the coefficients in the equation with the partial derivatives

$$\frac{\partial z}{\partial x} \quad \text{and} \quad \frac{\partial z}{\partial y}$$

233

For example, in the exact equation

$$(2x+3y)\,dx+(3x-2y)\,dy=0$$

$$\frac{\partial z}{\partial x}=2x+3y \tag{1}$$

and

$$\frac{\partial z}{\partial y}=3x-2y \tag{2}$$

It is now possible to suggest the function from which these two partial derivatives were formed.

From (1), z must contain the terms $x^2+3xy+C$

and from (2), z must contain the terms $3xy-y^2+C$

It must be concluded that the function will include all these terms without double counting any that are common to both. This leads to the solution

$$z=x^2+3xy-y^2+C=0$$

and it is easily checked that this is the solution to the differential equation.

This differential equation has homogeneous coefficients and could have been solved by the method of Section 9.2.2 with the same result but with a good deal more effort. It is, therefore, extremely useful to be able to recognise an exact differential equation when it occurs and to apply the above method to obtain a solution.

9.2.5. A General Method for First-Order Linear Equations

It is generally possible to make a non-exact differential equation into an exact one by multiplying by an *integrating factor*. For example,

let

$$\frac{dy}{dx}=x+y$$

then

$$\frac{dy}{dx}-y=x$$

The left-hand side of the equation is not exact and it is difficult to integrate as it stands. But if all the terms are multiplied by e^{-x} the equation becomes

$$e^{-x} \frac{dy}{dx} - e^{-x}y = e^{-x}x$$

The left-hand side is now exact and is the derivative of $e^{-x}y$ and the right-hand side of the equation can be integrated by parts (see Section 6.10).

$$\int xe^{-x}\,dx = x(-e^{-x}) - \int (-e^{-x})\,dx$$

$$= -xe^{-x} - e^{-x} + C = e^{-x}(-x-1) + C$$

This is equal to the integral of the left-hand side, ie the exact part, of the equation

∴ $$e^{-x}y = e^{-x}(-x-1) + C$$

Dividing throughout by e^{-x}, we have

$$y = -x - 1 + \frac{C}{e^{-x}} = -x - 1 + Ce^x$$

therefore the solution to the differential equation

$$\frac{dy}{dx} = x + y \qquad \text{is} \qquad y = Ce^x - x - 1$$

This can be checked by differentiation:

$$\frac{dy}{dx} = Ce^x - 1 \qquad \text{and} \qquad Ce^x = x + y + 1$$

∴ $$\frac{dy}{dx} = x + y + 1 - 1 = x + y$$

In general, the integrating factor is equal to e^{θ} where $\theta = \int P\,dx$ for all differential equations of the form

$$\frac{dy}{dx} + Py = Q$$

where P and Q are functions of x only, including the case where they are constants. This method is useful where both P and the product of the integrating factor and Q are reasonably easy to integrate.

Further example $$\frac{dy}{dx} + 4y = 2x$$

In this case $P = 4$ and $\int P\,dx = 4x$

Therefore the integrating factor is e^{4x}, and multiplying by this gives

$$e^{4x}\frac{dy}{dx} + 4e^{4x}y = 2xe^{4x}$$

The left-hand side is now the derivative of ye^{4x}, and integrating the right-hand side gives

$$\int 2xe^{4x}\,dx = \frac{2xe^{4x}}{4} - \int \frac{e^{4x}}{4}2\,dx = \frac{xe^{4x}}{2} - \frac{e^{4x}}{8} + A$$

$$\therefore \qquad ye^{4x} = \frac{xe^{4x}}{2} - \frac{e^{4x}}{8} + A \qquad \text{or} \qquad y = \frac{x}{2} - \frac{1}{8} + Ae^{-4x}$$

9.3 EXERCISES

1. Find the solution to the equation

$$(a_1 Y - b_1 X)\,dX - (a_2 Y - b_2 X)\,dY = 0.$$

2. What is the general form of the demand equation which has an elasticity of $-n$?

3. The total cost of production, y, and the level of output x are related to the marginal cost of production by the equation

$$\frac{dy}{dx} = \frac{24x^2 - y^2}{xy}$$

What is the total cost function if $x = 2$ when $y = 4$?

4. Determine the consumption function for which the marginal propensity to consume is

$$\frac{dc}{dx} = a - bx$$

where c is consumption and x is income.

5. With a Walrasian adjustment process, if the quantities supplied and demanded differ then the rate of change of prices is given by

$$\frac{dp}{dt} = kz$$

where t is time, z is excess demand and p is the deviation of price from its equilibrium level. Solve the differential equation for p when

(a) $z = ap$ (b) $z = ap + bp^2$

6. The Harrod–Domar growth model can be formulated as follows:

$$S = aY \tag{1}$$

$$I = b\frac{dY}{dt} \tag{2}$$

$$I = S \tag{3}$$

Where S is savings, Y is income and I is investment. Substitute from (1) and (2) into (3) and solve the differential equation for Y.

7. Solve the following equations

(a) $x\dfrac{dy}{dx} = x - y$ (b) $\dfrac{dy}{dx} = \dfrac{y+1}{2y+x}$

(c) $\dfrac{dy}{dx} = x(1+x) - 2y$ (d) $\dfrac{dy}{dx} + 4y = x^3$

9.4 SECOND-ORDER HOMOGENEOUS DIFFERENTIAL EQUATIONS

The general form of linear second-order differential equations with constant coefficients is

$$a\frac{d^2y}{dx^2} + b\frac{dy}{dx} + cy = f(x)$$

where a, b and c are constants. When $f(x)$ is equal to zero the equation is said to be *homogeneous* and one solution of the homogeneous equation

237

$$a\frac{d^2y}{dx^2} + b\frac{dy}{dx} + cy = 0$$

is of the form $y = e^{mx}$ where m is a constant. To determine m let $y = e^{mx}$.

Then
$$\frac{dy}{dx} = me^{mx} \qquad \frac{d^2y}{dx^2} = m^2e^{mx}$$

Substituting these values in the homogeneous equation, we obtain

$$am^2e^{mx} + bme^{mx} + ce^{mx} = 0$$

$$e^{mx}(am^2 + bm + c) = 0$$

∴
$$(am^2 + bm + c) = 0$$

This is a quadratic equation in the unknown m and the roots can be found by the usual method (see Section 3.5). Therefore

$$m = \frac{-b \pm \sqrt{(b^2 - 4ac)}}{2a}$$

or
$$m_1 = \frac{-b + \sqrt{(b^2 - 4ac)}}{2a} \qquad m_2 = \frac{-b - \sqrt{(b^2 - 4ac)}}{2a}$$

This suggests two possible alternatives for the solution to the differential equation:

$$y = e^{m_1 x} \qquad \text{and} \qquad y = e^{m_2 x}$$

But the solution to a second-order differential equation must contain two constants of integration. This is achieved by combining the two solutions in the form

$$y = k_1 e^{m_1 x} + k_2 e^{m_2 x}$$

where k_1 and k_2 are the two arbitrary constants. This value of y satisfies the differential equation as can be shown by substitution.

Example

$$\frac{d^2y}{dx^2} - \frac{5\,dy}{dx} + 6y = 0$$

Then if
$$y = e^{mx}$$

$$m^2 e^{mx} - 5m e^{mx} + 6 e^{mx} = 0$$

$$e^{mx}(m^2 - 5m + 6) = 0 \quad \text{or} \quad m^2 - 5m + 6 = 0$$

$\therefore m_1 = 2$ and $m_2 = 3$ and the solution is

$$y = k_1 e^{2x} + k_2 e^{3x}$$

Check:
$$\frac{dy}{dx} = 2k_1 e^{2x} + 3k_2 e^{3x}$$

$$\frac{d^2 y}{dx^2} = 4k_1 e^{2x} + 9k_2 e^{3x}$$

Hence

$$\frac{d^2 y}{dx^2} - 5\frac{dy}{dx} + 6y = 4k_1 e^{2x} + 9k_2 e^{3x} - 10k_1 e^{2x} - 15k_2 e^{3x} + 6k_1 e^{2x} + 6k_2 e^{3x}$$

and this is equal to 0.

The term e^{mx} will always be ignored after the function has been differentiated and it is not necessary to include it once we are familiar with its purpose. If it is ignored the above procedure can be conveniently summarised in the following set of rules.

1. From the differential equation form a new equation in which $d^2 y/dx^2$ is replaced by m^2, dy/dx is replaced by m and y is replaced by unity:

$$am^2 + bm + c = 0$$

This is known as the *auxiliary equation*.
2. Find the roots of this quadratic equation, m_1 and m_2.
3. Form the solution to the differential equation

$$y = k_1 e^{m_1 x} + k_2 e^{m_2 x}$$

The values of the constants k_1 and k_2 can be found if two sets of data are available about the system.

Returning to the particular differential equation discussed above, let us assume that when $x = 0$, $y = 1$ and when $x = 0$, $dy/dx = 4$. This information can be used to find the values of k_1 and k_2.

239

For
$$y = k_1 e^{2x} + k_2 e^{3x}$$

$$\frac{dy}{dx} = 2k_1 e^{2x} + 3k_2 e^{3x}$$

Substituting +'   these equations

$$1 = k_1 e^0 + k_2 e^0 = k_1 + k_2$$

$$4 = 2k_1 e^0 + 3k_2 e^0 = 2k_1 + 3k_2$$

The solution of this pair of simultaneous equations is

$$k_1 = -1 \qquad k_2 = 2$$

The solution to the differential equation

$$\frac{d^2 y}{dx^2} - \frac{5\,dy}{dx} + 6y = 0$$

with $x = 0, \qquad y = 1; \qquad x = 0, \qquad dy/dx = 4$

is given by

$$y = -1e^{2x} + 2e^{3x} = 2e^{3x} - e^{2x}$$

The above method must be modified when the roots of the auxiliary equation are equal, ie $m_1 = m_2 = m$. The solution is given by

$$y = (k_1 + k_2 x)e^{mx}$$

If this modification is not applied the result contains only one constant of integration and this could not form the complete solution to a second-order differential equation:

$$y = k_1 e^{mx} + k_2 e^{mx} = (k_1 + k_2)e^{mx} = ke^{mx}$$

The basic result does, however, hold when the roots of the auxiliary equation are complex. For example,

$$m_1 = a + bi \qquad m_2 = a - bi$$

In this case the solution is given by

$$y = k_1 e^{(a+bi)x} + k_2 e^{(a-bi)x}$$

This can be expressed in a more convenient form if the following expansions are used (see Section 5.12):

$$e^{ix} = \cos x + i \sin x \qquad e^{-ix} = \cos x - i \sin x$$

Replacing x by bx gives

$$e^{bix} = \cos bx + i \sin bx \qquad e^{-bix} = \cos bx - i \sin bx$$

By algebraic manipulation

$$y = k_1 e^{(a+bi)x} + k_2 e^{(a-bi)x}$$

becomes $\quad y = k_1 e^{ax} e^{bix} + k_2 e^{ax} e^{-bix} = e^{ax}[k_1 e^{bix} + k_2 e^{-bix}]$

$$= e^{ax}[k_1(\cos bx + i \sin bx) + k_2(\cos bx - i \sin bx)]$$

and by collecting together the terms in $\cos bx$ and $\sin bx$

$$y = e^{ax}[(k_1 + k_2) \cos bx + i(k_1 - k_2) \sin bx]$$

$$= e^{ax}[k_3 \cos bx + k_4 \sin bx]$$

where $\qquad k_3 = k_1 + k_2 \qquad$ and $\qquad k_4 = i(k_1 - k_2)$

An even more useful form can be obtained by considering a right-angled triangle with sides equal to k_3, k_4 and $\sqrt{(k_3^2 + k_4^2)}$ as shown in Fig 9.1. Let the angle indicated be ϵ.

Then $\qquad \cos \epsilon = \dfrac{k_3}{\sqrt{(k_3^2 + k_4^2)}} \qquad$ and $\qquad \sin \epsilon = \dfrac{k_4}{\sqrt{(k_3^2 + k_4^2)}}$.

These results can be used if the solution to the differential equation is written in a slightly different form.

$$y = e^{ax}[k_3 \cos bx + k_4 \sin bx]$$

$$= e^{ax}\sqrt{(k_3^2 + k_4^2)} \left[\cos bx \, \frac{k_3}{\sqrt{(k_3^2 + k_4^2)}} + \sin bx \, \frac{k_4}{\sqrt{(k_3^2 + k_4^2)}} \right]$$

$$= e^{ax}\sqrt{(k_3^2 + k_4^2)}[\cos bx \cos \epsilon + \sin bx \sin \epsilon]$$

$$= e^{ax}A[\cos bx \cos \epsilon + \sin bx \sin \epsilon]$$

where $A = \sqrt{(k_3^2 + k_4^2)}$ and is a constant.

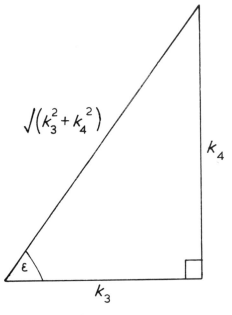

Fig 9.1

Using the trigonometrical identity

$$\cos (bx - \epsilon) = \cos bx \cos \epsilon + \sin bx \sin \epsilon$$

(see Appendix A, Section 2) the solution can now be written in the form

$$y = A e^{ax} \cos (bx - \epsilon)$$

where A and ϵ are the two necessary constants with values

$$A = \sqrt{(k_3^2 + k_4^2)} \qquad \text{and} \qquad \epsilon = \tan^{-1} (k_4/k_3)$$

The values of both can be obtained from the initial data about the system. Thus the solution to a second-order homogeneous differential equation whose auxiliary equation has complex roots involves trigonometrical functions.

Examples The relationship between national income, y, and time, x, is given by

(a) $\dfrac{d^2y}{dx^2} - 3\dfrac{dy}{dx} + 2y = 0$

(b) $\dfrac{d^2y}{dx^2} - 10\dfrac{dy}{dx} + 25y = 0$

(c) $\dfrac{d^2y}{dx^2} + 6\dfrac{dy}{dx} + 10y = 0$

In each case, obtain y as a function of x if $x = 0$ when $y = 0$ and $dy/dx = 2$ when $x = 0$.

(a) $\dfrac{d^2y}{dx^2} - 3\dfrac{dy}{dx} + 2y = 0$.

Let $y = e^{mx}$. Then

$$\frac{dy}{dx} = me^{mx} \qquad \text{and} \qquad \frac{d^2y}{dx^2} = m^2 e^{mx}$$

Substituting in the equation gives

$$e^{mx}(m^2 - 3m + 2) = 0$$

and so

$$m = \frac{+3 \pm \sqrt{(9-8)}}{2} = 2 \text{ or } 1.$$

$\therefore$ $\qquad\qquad\qquad y = Ae^{2x} + Be^{x}$

The initial conditions are $x = 0$, $y = 0$ and $x = 0$, $dy/dx = 2$. Substituting gives two equations in A and B:

$$0 = A + B$$

and since $dy/dx = 2Ae^{2x} + Be^{x}$, $2 = 2A + B$. Therefore, solving for A and B gives $A = -B$ and so $A = 2$, $B = -2$.

$\therefore$ $\qquad\qquad\qquad y = 2e^{2x} - 2e^{x}$

(b) $\dfrac{d^2y}{dx^2} - 10\dfrac{dy}{dx} + 25y = 0$

Let $y = e^{mx}$. Then

$$\frac{dy}{dx} = me^{mx} \qquad \text{and} \qquad \frac{d^2y}{dx^2} = m^2 e^{mx}$$

Substituting in the equation gives

$$e^{mx}(m^2 - 10m + 25) = 0$$

and so $\qquad m = \dfrac{10 \pm \sqrt{(100 - 100)}}{2} = 5$

$\therefore \qquad y = (A + Bx)e^{5x}$

$$\frac{dy}{dx} = (A + Bx)5e^{5x} + Be^{5x} = (5A + B + 5Bx)e^{5x}$$

When $\qquad x = 0, \qquad y = 0 \qquad$ and $\qquad A = 0$

When $\qquad x = 0, \qquad dy/dx = 2, \qquad$ so $\qquad 2 = 5A + B$

$\therefore \qquad A = 0$ and $B = 2$, and the solution is $y = 2xe^{5x}$

(c) $\dfrac{d^2y}{dx^2} + 6\dfrac{dy}{dx} + 10y = 0$

Let $y = e^{mx}$. Then

$$\frac{dy}{dx} = me^{mx} \qquad \text{and} \qquad \frac{d^2y}{dx^2} = m^2 e^{mx}$$

Substituting into the equation gives

$$e^{mx}(m^2 + 6m + 10) = 0$$

and so $\qquad m = \dfrac{-6 \pm \sqrt{(36 - 40)}}{2} = -3 \pm i$

We know that if $m = a \pm bi$ then

$$y = Ae^{ax} \cos(bx - \epsilon)$$

and so since $m = -3 \pm i$, $a = -3$ and $b = 1$.

$\therefore \qquad y = Ae^{-3x} \cos(x - \epsilon)$

The initial conditions are $x = 0$, $y = 0$ which gives

$$0 = A \cos (-\epsilon) \tag{1}$$

and $$x = 0, \quad \frac{dy}{dx} = 2$$

Now $$\frac{dy}{dx} = -3Ae^{-3x} \cos (x - \epsilon) - Ae^{-3x} \sin (x - \epsilon)$$

$$\therefore \qquad 2 = -3A \cos (-\epsilon) - A \sin (-\epsilon) \tag{2}$$

From (1) either $A = 0$ or $\cos (-\epsilon) = 0$, that is $\epsilon = \pi/2$

From (2): $\qquad 2 = -A[3(0) + \sin (-\pi/2)] = -A(-1)$

$$\therefore \qquad A = +2, \qquad \epsilon = \pi/2 \qquad \text{and} \qquad y = 2e^{-3x} \cos (x - \pi/2)$$

9.5 EXERCISES

Solve the following differential equations:

1. $\dfrac{d^2y}{dx^2} - \dfrac{dy}{dx} - 12y = 0$

2. $\dfrac{d^2y}{dx^2} - 3\dfrac{dy}{dx} + 2y = 0$

3. $\dfrac{d^2y}{dx^2} - 6\dfrac{dy}{dx} + 9y = 0$

4. $\dfrac{d^2y}{dx^2} - 2\dfrac{dy}{dx} + y = 0$

5. $\dfrac{d^2y}{dx^2} - 16y = 0$

6. $\dfrac{d^2y}{dx^2} + 16y = 0$

7. $\dfrac{d^2y}{dx^2} + 2\dfrac{dy}{dx} + 10y = 0$

9.6 SECOND-ORDER NON-HOMOGENEOUS DIFFERENTIAL EQUATIONS

The *non-homogeneous* second-order differential equation with constant coefficients has the general form

$$a \frac{d^2y}{dx^2} + b \frac{dy}{dx} + cy = f(x)$$

where $f(x)$ is either a constant or a function of x only, and a, b, c are constants.

It can be shown that the solution to this type of equation is the sum of two parts which are obtained independently. These are called the *complementary function* (CI) and the *particular integral* (PI). The complementary function is simply the solution of the homogeneous part of the equation formed by letting $f(x) = 0$ and this can be obtained by the method of Section 9.4. The particular integral is any solution which satisfies the differential equation and its form depends upon the form of $f(x)$.

(a) $f(x)$ is a constant. For example,

$$\frac{d^2y}{dx^2} - 5 \frac{dy}{dx} + 6y = 18$$

If we let $y = K = $ a constant then

$$\frac{dy}{dx} = 0, \qquad \frac{d^2y}{dx^2} = 0$$

Substituting these values in the differential equation

$$0 - 0 + 6K = 18 \qquad \text{or} \qquad K = 3$$

Thus the solution $y = 3$ is a particular integral because it satisfies the differential equation.

The complementary function was obtained in Section 9.4 when the solution to

$$\frac{d^2y}{dx^2} - 5 \frac{dy}{dx} + 6y = 0$$

was shown to be $y = k_1 e^{2x} + k_2 e^{3x}$. Therefore the complete solution to the differential equation

$$\frac{d^2y}{dx^2} - 5\frac{dy}{dx} + 6y = 18$$

is the sum of the complementary function and the particular integral and is given by

$$y = k_1 e^{2x} + k_2 e^{3x} + 3$$

This can be verified by differentiation and substitution in the usual way.

(b) $f(x)$ is linear in x. For example,

$$\frac{d^2y}{dx^2} - 5\frac{dy}{dx} + 6y = 18x$$

Let us try a solution of the form $y = K_1 x + K_2$. Then

$$\frac{dy}{dx} = K_1, \qquad \frac{d^2y}{dx^2} = 0$$

Substituting these values in the differential equation

$$0 - 5K_1 + 6(K_1 x + K_2) = 18x$$

or

$$6K_1 x - 5K_1 + 6K_2 = 18x$$

For this equation to be satisfied it must be true for all values of x

$\therefore$ $$6K_1 = 18$$

and $$-5K_1 + 6K_2 = 0$$

$\therefore$ $$K_1 = 3 \qquad \text{and} \qquad K_2 = \tfrac{5}{2}$$

The particular integral is then

$$y = 3x + \tfrac{5}{2}$$

and the complete solution to the differential equation

$$\frac{d^2y}{dx^2} - 5\frac{dy}{dx} + 6y = 18x$$

is $$y = k_1 e^{2x} + k_2 e^{3x} + 3x + \tfrac{5}{2}$$

247

TABLE 9.1

$f(x)$	Particular integral
$ax^2 + bx + c$	$K_1 x^2 + K_2 x + K_3$
e^x	$K_1 e^x$
$\sin x$	$K_1 \sin x + K_2 \cos x$

From the above results it would seem reasonable to expect the particular integral to assume the form that $f(x)$ has in the differential equation and this is indeed so in simple cases. Table 9.1 gives a few examples. The reason that the last of these functions contains both sine and cosine terms is that the differentiation of one leads to the other and therefore a combination of two terms is used.

For example
$$\frac{d^2 y}{dx^2} - 5 \frac{dy}{dx} + 6y = 18 \sin x$$

Let the particular integral be represented by
$$y = K_1 \sin x + K_2 \cos x$$

then
$$\frac{dy}{dx} = K_1 \cos x - K_2 \sin x$$

$$\frac{d^2 y}{dx^2} = -K_1 \sin x - K_2 \cos x$$

Substituting these values in the differential equation, we obtain

$$(-K_1 \sin x - K_2 \cos x) - 5(K_1 \cos x - K_2 \sin x)$$
$$+ 6(K_1 \sin x + K_2 \cos x) = 18 \sin x$$

Equating the coefficients of the sine and cosine terms

$$\sin x(-K_1 + 5K_2 + 6K_1) = 18 \sin x$$
$$\cos x(-K_2 - 5K_1 + 6K_2) = 0$$

that is
$$5K_1 + 5K_2 = 18$$
$$-5K_1 + 5K_2 = 0$$

248

From which $K_2 = \frac{9}{5}$ and $K_1 = \frac{9}{5}$

∴ The solution to the differential equation

$$\frac{d^2y}{dx^2} - 5\frac{dy}{dx} + 6y = 18 \sin x$$

is $\qquad y = k_1e^{2x} + k_2e^{3x} + \frac{9}{5}(\sin x + \cos x)$

In some cases the particular integral may not be easy to find using the above procedure.

For example, difficulty is experienced in the case where the term in y is not present in the left-hand side of the equation and d is a constant. For example,

$$\frac{d^2y}{dx^2} - 3\frac{dy}{dx} = 16$$

The auxiliary equation is

$$m^2 - 3m = 0$$

that is $\qquad m(m-3) = 0$

∴ $\qquad m_1 = 0$ and $m_2 = 3$

and $\qquad CF = k_1e^{0x} + k_2e^{3x} = k_1 + k_2e^{3x}$

For the particular integral let $y = K$. Then

$$\frac{dy}{dx} = 0, \qquad \frac{d^2y}{dx^2} = 0$$

and substitution of these values leads to $0 - 0 = 6$, which is an impossible result.

In such a case we can try $y = Kx$. Then

$$\frac{dy}{dx} = K, \qquad \frac{d^2y}{dx^2} = 0$$

and on substitution $\qquad 0 - 3K = 16$

and $\qquad K = -\frac{16}{3}$

and $\qquad PI = -\frac{16}{3}x$

The solution to the differential equation

$$\frac{d^2y}{dx^2} - 3\frac{dy}{dx} = 16$$

is

$$y = k_1 + k_2 e^{3x} - \tfrac{16}{3}x$$

In general, the particular integral can be found by this trial and error method. It is also possible, and sometimes better, to find it by the method of operators. This is not, however, discussed here because very little advantage is gained by its use in simple cases. The reader should consult a specialised text for a discussion of the method and its application to more difficult problems.

9.7 EXERCISES

1. Solve the following differential equations with the given initial conditions:

(a) $\dfrac{d^2y}{dx^2} - 10\dfrac{dy}{dx} + 16y = 2x$ with $x=0$, $y = -5/64$ and

$x=0$, $\dfrac{dy}{dx} = 1$

(b) $\dfrac{d^2y}{dx^2} - 16y = e^x$ with $x=0$, $y = 14/15$ and $x=0$,

$\dfrac{dy}{dx} = 29/15$

(c) $\dfrac{d^2y}{dx^2} - 2\dfrac{dy}{dx} + y = e^{2x}$ with $x=0$, $y=2$ and $x=1$, $y=e^2$

(d) $\dfrac{d^2y}{dx^2} + 2\dfrac{dy}{dx} + 5y = x$ with $x=0$, $y = -2/25$ and $x=0$,

$\dfrac{dy}{dx} = 6/5$

(e) $\dfrac{d^2y}{dx^2} - \dfrac{dy}{dx} = 2\cos x$ with $x=0$, $y=1$ and $x=0$,

$dy/dx = 2$

2. The relationship between output (x), and total cost (y) for a firm is given by

$$\frac{d^2y}{dx^2} + \frac{dy}{dx} - 6y = 18x^2$$

where dy/dx is the marginal cost, and d^2y/dx^2 is the rate of change of marginal cost. Given that $y=0$ and $dy/dx=1$ when $x=0$, obtain the total cost as a function of output only.

9.8 GRAPHICAL PRESENTATION

The solution of a differential equation expresses one of the variables as a function of the other. In many economic problems the independent variable is time and it is of particular interest to consider the variations which occur in the dependent variable over a period of time. A convenient way of presenting this information is in the form of a graph with the time axis horizontal. It is impossible to discuss all the types of differential equations, but a few examples are given here to indicate the method to be used in obtaining a graph of the solution in any particular case.

LINEAR FIRST-ORDER DIFFERENTIAL EQUATIONS

The solution to an equation of this form is

1. a polynomial in t. The graph of this must be obtained by plotting a few values of the variables. No general form exists.

2. an exponential function $y = ke^{at}$. In this case various possibilities exist.

    (a) k is positive and a is positive (Fig 9.2) e^{at} continuously increases with time and at an ever-increasing rate. Therefore the function becomes very large and is said to be explosive. The value of k determines the starting position, ie the value of the function at time $t=0$.

    (b) k is positive and a is negative (Fig 9.3). This function continuously decreases with time and is said to be damped.

    (c) k is negative and a is positive.

    (d) k is negative and a is negative.

251

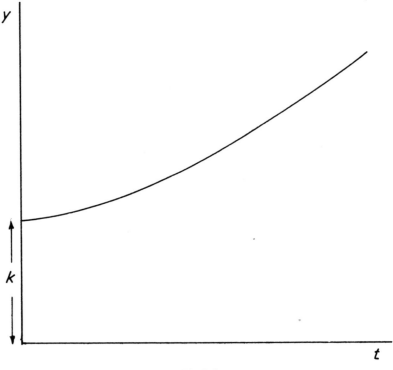

Fig 9.2

The last two functions are the mirror images in the time axis of the functions which are sketched in Figs 9.2 and 9.3.

LINEAR SECOND-ORDER DIFFERENTIAL EQUATIONS WITH CONSTANT COEFFICIENTS

The general equation

$$a \frac{d^2y}{dt^2} + b \frac{dy}{dt} + cy = f(x)$$

has a solution of the form

$$y = k_1 e^{m_1 t} + k_2 e^{m_2 t} + PI$$

252

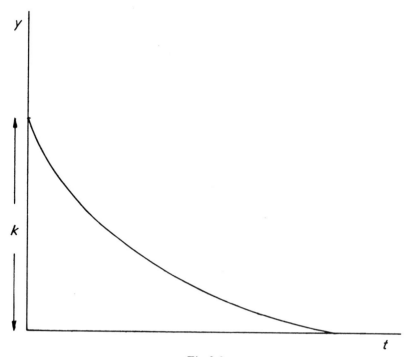

Fig 9.3

where PI depends upon the form of $f(x)$.

1. The complementary function is the sum of two exponential terms and the graph will depend upon the values of m_1 and m_2, the roots of the auxiliary equation.

(a) $m_1 > 0$ $m_2 > 0$, $k_1 > 0$ and $k_2 > 0$

The function will increase continuously from the value $y = k_1 + k_2$ at time $t = 0$ and will, for any value of t, be equal to the sum of the two separate functions

$$y = k_1 e^{m_1 t} \quad \text{and} \quad y = k_2 e^{m_2 t}$$

If m_1 is very much greater than m_2 (written $m_1 \gg m_2$) then $k_1 e^{m_1 t} \gg k_2 e^{m_2 t}$ as t becomes very large and the function approximates to $y = k_1 e^{m_1 t}$.

(b) $m_1 < 0,$ $\quad m_2 < 0,$ $\quad k_1 > 0$ $\quad$ and $\quad k_2 > 0$

The function decreases continuously over time from the value $(k_1 + k_2)$ and approaches the value zero.

(c) $m_1 > 0,$ $\quad m_2 < 0,$ $\quad k_1 > 0$ $\quad$ and $\quad k_2 > 0$

The term $y = k_2 e^{m_2 t}$ decreases continuously as in (b) above because m_2 is negative and the function approaches the function $y = k_1 e^{m_1 t}$ as t increases.

The same reasoning can be applied to other possible combinations of the constants in the complementary function and the shape of the function determined.

(d) Complex roots

The solution is of the form

$$y = A e^{at} \cos (kt - \epsilon)$$

and this can be thought of as two parts. Ae^{at} has a form similar to one of those discussed above, and $\cos (kt - \epsilon)$ is a function which can take on all values between -1 and $+1$ but has no values outside these limits. This means that the solution oscillates about the time axis with extreme values Ae^{at} and $-Ae^{at}$ (Fig 9.4).

The amplitude of the oscillations increases at each successive cycle and the function is explosive. The constant ϵ helps to determine the starting point in the cycle because when $t = 0$

$$y = A \cos (-\epsilon) = A \cos \epsilon$$

If $\epsilon = 0$ the function has an initial value which is determined by the value of A. For all other values of ϵ the initial value of the function is less than A with the extreme case of $\cos \epsilon = -1$ when the function starts at $-A$.

ϵ is said to determine the *phase*, or position, within the cycle at which the function starts.

2. The effect of the particular integral is to modify the function in a manner which is determined by the form of the integral.

(a) *PI* is a constant

The shape of the function is unaltered but its value at time

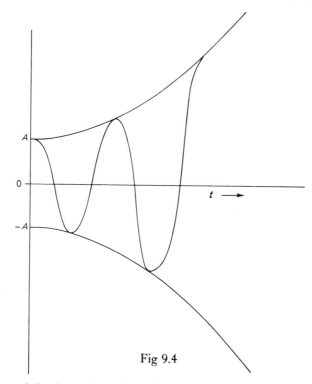

Fig 9.4

$t = 0$ is changed to the value of the constant. That is, when $t = 0$, $y = k_1 + k_2 + c$.

(b) *PI* is a polynomial in t

This can have a large effect upon the shape of the curve particularly for small values of t. It has little effect for larger values of t, on functions which contain exponential terms which are explosive.

(c) *PI* is a trigonometric function

This will add a periodic oscillation to the function, the amplitude and period of which will be determined by the trigonometric function.

9.9 EXERCISES

Sketch the graphs of the solutions to Exercises 9.7.

CHAPTER TEN

Difference Equations

10.1 INTRODUCTION

In Chapter 9 we saw that differential equations express the relationship between two variables (eg x and y) and also the rate of change of one variable with respect to the other, (ie dy/dx). We know that this rate of change (or derivative) relates to variables which change continuously. In this chapter we will be considering variables which change in discrete steps and in particular variables which are measured at or over certain periods of time. Examples of such variables are the level of stocks at the end of a month, the national income earned during a calendar year, the daily closing price of shares on the stock exchange, and the value of a bank deposit at the end of a year.

For example, if $£C$ is invested in a bank at an interest rate of i per cent per year compounded annually, then the value of this investment after one year, C_1, is given by

$$C_1 = (1+i)C$$

and the value after two years, C_2, is

$$C_2 = (1+i)C_1 = (1+i)^2 C$$

In general, the value of the investment at the end of year t is

$$C_t = (1+i)^t C = a^t C \qquad \text{where} \qquad a = 1+i$$

Now
$$C_t = (1+i)C_{t-1} = aC_{t-1}$$

where a relates the value of the investment after t years to its value after $(t-1)$ years. This expression is known as a *difference equation*.

It is a *first-order* difference equation because the time periods

256

involved are only one period apart. The value in any time period can be determined if the value in the preceding time period is known.

Thus
$$C_t = aC_{t-1}$$
$$C_{t+1} = aC_t = a^2C_{t-1}$$
$$C_{t+2} = aC_{t+1} = a^3C_{t-1}$$

In particular, if $C_0 = C$, then $C_t = a^t C$.

This equation is said to be the *solution* to the difference equation $C_t = aC_{t-1}$ since it enables us to determine the value of the investment at any time period directly from its value initially.

For example, if $C = £100$ and the rate of interest is 5 per cent per annum then $a = 1 + 0.05 = 1.05$ and $C_t = (1.05)^t \times 100$ so that after 20 years, $C_{20} = 100(1.05)^{20} = 265.33$.

In general, the necessary requirements for a solution to a difference equation are as follows.

1. It expresses the value of the variable in time period t in terms of the given data, ie the value of the variable in the base period. It can therefore be used to determine the value of the variable in any time period directly without calculating all the intermediate values as would be necessary if the original difference equation were used.
2. It satisfies the original difference equation.

For the example used above, the second condition can be checked as follows:

Let $C_t = a^t C_0$ be the solution to the difference equation $C_t = aC_{t-1}$; then $C_{t+1} = a^{t+1}C_0$ is obtained by substituting $t+1$ for t whenever it occurs in the solution. Now

$$C_{t+1} = aC_t = a(a^t C_0) = a^{t+1}C_0$$

and so $C_t = a^t C_0$ satisfies the original difference equation.

10.2 COMPOUND INTEREST AND THE ADDITION OF CAPITAL AT YEARLY INTERVALS

It is possible to construct a difference equation to represent the situation in which the capital sum has interest added to it annually along with a further injection of capital. For example, an initial capital sum C with interest i per cent compounded annually has a further sum of A added at the end of each year.

Then with $a = 1 + i$ as above,

$$C_0 = C$$
$$C_1 = C_0(1+i) + A = aC_0 + A$$
$$C_2 = (aC_0 + A)(1+i) + A = a^2C_0 + aA + A$$
$$C_3 = (a^2C_0 + aA + A)(1+i) + A = a^3C_0 + a^2A + aA + A$$

and in general,

$$C_t = a^tC_0 + A(a^{t-1} + a^{t-2} + \cdots + 1)$$

The first term on the right-hand side of the equation is identical to that in the previous example where A was not added at the end of each year. The second term is a geometric progression the sum of which is quite easily obtained (see Section 4.4) and the result is

$$C_t = a^tC_0 + A\left(\frac{a^t - 1}{a - 1}\right)$$

This is a solution because

(a) it enables us to determine the value of the capital sum at any time period from a knowledge of C, a and A, the three given quantities;

(b) it satisfies the difference equation

$$C_{t+1} = C_t(1+i) + A$$
$$= aC_t + A$$

For example, let £100 be invested at an interest rate of 10 per cent per year compounded annually and let the capital added each year be £20.

Then
$$C_0 = 100$$
$$C_1 = 100(1+0.10) + 20$$
$$C_2 = 100(1+0.10)^2 + 20(1+0.10) + 20$$
$$\vdots$$

The value after any time period can be determined by considering the solution

$$C_t = a^t C_0 + A \left(\frac{a^t - 1}{a - 1} \right)$$

where

$$a = 1 + i = 1 + 0.10 = 1.1$$

then

$$C_t = 1.1^t(100) + 20 \left(\frac{1.1^t - 1}{1.1 - 1} \right)$$

$$= 100(1.1)^t + 20 \left(\frac{(1.1)^t - 1}{0.1} \right)$$

$$= 100(1.1)^t + 200(1.1^t - 1)$$

After 5 years, $t = 5$ and

$$C_5 = 100(1.1)^5 + 200[(1.1)^5 - 1]$$

$$= 100(1.61) + 200(1.61 - 1) = 161 + 200(0.61)$$

$$= 161 + 122 = 283$$

The solution in this case is made up of two terms, the second of which arises from the presence of the additional term in the difference equation.

There are many other economic examples of situations which can be conveniently represented by difference equations. It is possible to develop solutions to these from first principles in the above manner but this is not really necessary because a general theory of difference equations is available which will enable us to arrive quickly and easily at the correct solution.

This is best explained and discussed if we classify difference equations into types, in the manner which was adopted for differential equations in Chapter 9.

10.3 FIRST-ORDER LINEAR DIFFERENCE EQUATIONS

10.3.1. Homogeneous

These can be written in general

$$Y_{t+1} = \lambda Y_t$$

where λ is a constant and the solution is given by

$$Y_t = k(\lambda)^t$$

k is a constant whose value is determined by the initial conditions. For example, if $Y_t = C$ when $t = 0$, then $C = k\lambda^0 = k$ and the solution is $Y_t = C(\lambda)^t$. As with differential equations, the solution to every first-order difference equation contains one arbitrary constant whose value can be determined only from other information about the system.

10.3.2. NON-HOMOGENEOUS

These can be written in general as

$$Y_{t+1} = \lambda Y_t + f(t)$$

or

$$Y_{t+1} - \lambda Y_t = f(t)$$

The solution to such an equation is obtained in two parts. The first is the solution of the homogeneous form $Y_{t+1} = \lambda Y_t$ and this is obtained as in Section 10.3.1. It is often known as the *transient solution*. The second part is a *particular solution* to the difference equation and its form depends upon the form of $f(t)$. It is often known as the *equilibrium solution*. The complete solution is the sum of these two parts.

When $f(t) = K$, where K is a constant, the equilibrium solution is, from Section 10.2:

$$Y_t = K \frac{1 - \lambda^t}{1 - \lambda}$$

The complete solution to the difference equation

$$Y_{t+1} - \lambda Y_t = K$$

is

$$Y_t = Y_0(\lambda)^t + K \frac{1 - \lambda^t}{1 - \lambda} = A\lambda^t + \frac{K}{1 - \lambda}$$

where A is a constant.

First-order difference equations in which $f(t)$ is not constant will be considered along with second-order equations in which the same condition holds.

10.4 THE COBWEB MODEL

In Section 1.7, we discussed the situation in which the quantity demanded and supplied of a commodity are equated under conditions of perfect competition. Thus,

$$q_D = f(p) \qquad \text{demand function}$$
$$q_s = f(p) \qquad \text{supply function}$$

and

$$q_D = q_s$$

But in many cases the response to changes in demand is not instantaneous. Thus a price increase does not immediately result in an increased supply. This is particularly so of such things as agricultural products, the supply of which cannot be altered to any great extent in a very short time. The rate of change in the quantity which can be supplied is determined by the time cycle of growth or production of the good. Once a crop is planted its size is governed by weather conditions, etc, and an increase in price during the growth and harvesting period has little effect on this year's supply. Similarly industries which are working at full capacity cannot increase the supply without putting in hand a capital-investment programme which is likely to take some time to implement. This problem can, of course, be overcome to some extent by allowing goods to be transferred into and out of inventory, ie by having a buffer stock in which fluctuations can take place instantaneously. If, initially, this possibility is ignored it seems reasonable to assume that a change in the price which can be obtained for a good has a delayed effect upon the quantity of the good supplied.

The simplest case we can consider is one in which the delay, or lag is for a single time period and applies to the supply function only. This can be written as follows:

$$q_t = f_D(p_t) \qquad \text{demand function}$$
$$q_t = f_s(p_{t-1}) \qquad \text{supply function}$$

This model states that the price in any time period determines the quantity supplied in the subsequent time period and this latter quantity when considered in relation to the demand function determines the price in that period. This price then determines the quantity supplied in the next period and so on.

261

If the price in two successive periods is different, then the quantity supplied in these two periods is different, and it is not difficult to visualise a situation where the price and the supply continue to change from period to period. If this occurs it is of interest to see whether there is an equilibrium position for the system, and to ask how the price and quantity supplied changes over a period of time. The situation can be represented graphically and the time path of the system determined.

An example where the demand and supply functions are both linear is illustrated in Fig 10.1. Given that the system starts with a price p_1 in the initial time period, the supply function enables us to determine the quantity q_2 which will be supplied at this price in the next time period. But if q_2 is supplied in the period 2 it can demand a price p_2 which is considerably higher than that in the preceding period. This figure can be found on the graph by moving vertically from the supply function to the demand function.

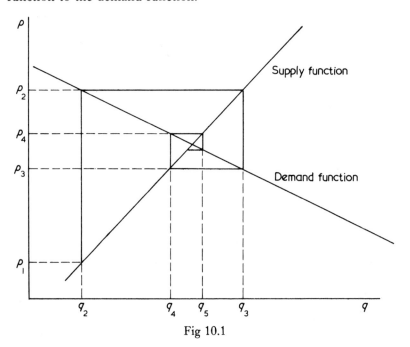

Fig 10.1

This high price results in an increase in the quantity supplied in the following time period. In this case it is equal to q_3. This figure can be found on the graph by moving horizontally from the demand function to the supply function.

From this point onwards, the procedure is to repeat the above steps, ie moving alternately from the supply to the demand function in vertical steps and from the demand to the supply function in horizontal steps. The result is that the changes in price and quantity between successive time periods become smaller and smaller and the system approaches an equilibrium value which is located at the intersection of the demand and supply functions.

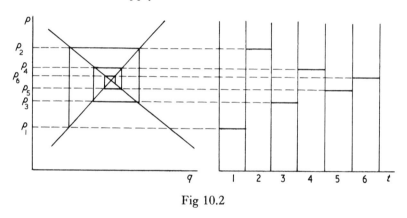

Fig 10.2

It is interesting to show the changes in price which occur by transferring the information to a graph where time is represented on the horizontal axis (Fig 10.2). The price fluctuates between time periods, the size of the fluctuations decreasing with time. The system tends to an equilibrium value and is, therefore, said to be *stable* or *converging*.

Figure 10.3 illustrates a situation in which the size of the fluctuations increases with time. This system is said to be *unstable, diverging* or *explosive*, and if it remained unchanged (which is highly unlikely) it would oscillate violently and the price of the good would fluctuate wildly from year to year.

A third situation is possible in which the price changes between successive time periods but alternates in such a way that it assumes

263

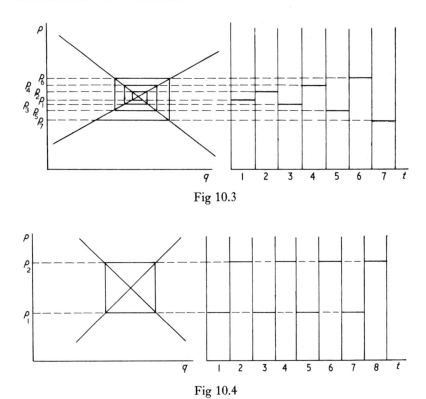

Fig 10.3

Fig 10.4

either one or other of only two values, ie the system *oscillates* (Fig 10.4).

Any linear system conforms to one of these three types. The particular type is determined by the slopes of the demand and supply functions and it is possible to decide into which category a system falls by following the above procedure, ie by constructing the time path of price from the demand and supply functions. But this is not really necessary in the linear case, because the same information can be obtained from the equations of these functions.

For example, let

$$q_t = f_D(p_t) = a_1 + b_1 p_t$$

264

and
$$q_t = f_s(p_{t-1}) = a_2 + b_2 p_{t-1}$$

Then if we assume that the quantity supplied equals the quantity demanded in any time period, it is possible to write

$$f_s(p_{t-1}) = f_D(p_t)$$

that is
$$a_2 + b_2 p_{t-1} = a_1 + b_1 p_t$$

This is a first-order difference equation

$$b_1 p_t = b_2 p_{t-1} + a_2 - a_1$$

or
$$p_t = \left[\frac{b_2}{b_1}\right] p_{t-1} + \frac{a_2 - a_1}{b_1}$$

The solution to the homogeneous form

$$p_t = \frac{b_2}{b_1} p_{t-1} \qquad \text{is given by} \qquad p_t = k \left(\frac{b_2}{b_1}\right)^t$$

(see Section 10.3.1) and it can be shown that the particular solution is (see Section 10.3.2)

$$p_t = \frac{(a_2 - a_1)/b_1}{1 - (b_2/b_1)} = \frac{a_2 - a_1}{b_1 - b_2}$$

The complete solution is therefore,

$$p_t = k \left[\frac{b_2}{b_1}\right]^t + \frac{a_2 - a_1}{b_1 - b_2}$$

where k is an arbitrary constant, the value of which is determined from the initial conditions.

The particular solution is a constant and has no influence on the changes which occur in the time path of price. These are determined by the solution to the homogeneous form. The manner in which the price changes between successive time periods is determined by the term b_2/b_1, where b_1 is the slope of the demand curve and is a function of the elasticity of demand which can be assumed, in general, to be negative. Similarly b_2 is a function of the elasticity of supply which is positive. The term b_2/b_1 must, therefore, be negative, $(b_2/b_1)^2$ is positive and $(b_2/b_1)^3$ is negative and so on. Its value is alternately positive and negative in each successive time period. It follows from the fact that the particular solution is constant that the function will oscillate about the value $p_t = (a_2 - a_1)/(b_1 - b_2)$ and the size of the oscillation depends upon the ratio b_2/b_1. Three distinct cases arise.

1. $|b_2/b_1| < 1$, or $|b_2| < |b_1|$, that is the modulus or absolute value of b_2 is less than that of b_1. In this case $(b_2/b_1)^t$ decreases in value as t increases and the size of the oscillations decreases over time. This corresponds to the stable system and the equilibrium price is $(a_2 - a_1)/(b_1 - b_2)$.

2. $|b_2/b_1| = 1$, that is $|b_2| = |b_1|$. In this case $(b_2/b_1)^t$ has an absolute value of unity but is alternately positive and negative as t changes. This corresponds to the situation in which the price takes on only two values

$$\frac{a_2 - a_1}{b_1 - b_2} + k \quad \text{and} \quad \frac{a_2 - a_1}{b_1 - b_2} - k$$

3. $|b_2/b_1| > 1$, that is $|b_2| > |b_1|$.

In this case $(b_2/b_1)^t$ increases in absolute value as t increases, changing sign between successive time periods. This corresponds to the unstable system because the size of the oscillations becomes larger and larger.

It is apparent, from this analysis, that the stability or otherwise of any demand–supply system depends upon the relative slopes of the two functions. In most practical situations these functions are not linear and therefore the slopes will be functions of price. In such cases it is possible to visualise a system which is unstable over one price range but stable for changes which are outside this price range. This can be demonstrated by the graphical approach and the result confirmed algebraically.

10.5 THE HARROD–DOMAR GROWTH MODEL

This assumes that

1. The amount of resources saved by the community during any time period is a function only of the income received by the community during that period.

That is
$$S_t = aY_t$$

where S_t represents savings in time period t

Y_t represents income in time period t

a is a positive constant which is less than 1

2. The amount of investment which is desirable in any time period will be a function of the increase of income in that period over the income of the previous time period.

That is
$$I_t = g(Y_t - Y_{t-1})$$

where g is a constant which is less than 1.

3. The amount which will be available for investment in any time period will be equal to the amount saved in the same time period.

That is
$$I_t = S_t$$

Using these three assumptions it is possible to construct a model of the system as follows.

$$S_t = a Y_t \tag{1}$$

$$I_t = g(Y_t - Y_{t-1}) \tag{2}$$

$$I_t = S_t \tag{3}$$

If eqs. (1) and (2) are substituted into eq. (3) a first-order linear difference equation is formed.

$$g(Y_t - Y_{t-1}) = a Y_t$$

$$Y_t(g - a) = g Y_{t-1}$$

$$Y_t = \left(\frac{g}{g-a} \right) Y_{t-1}$$

The solution to this is

$$Y_t = k \left(\frac{g}{g-a} \right)^t = k \lambda^t$$

where k is equal to the value of income in time period $t = 0$

Since
$$Y_0 = k \left(\frac{g}{g-a} \right)^0 = k$$

267

then
$$Y_t = Y_0 \left(\frac{g}{g-a} \right)^t$$

From assumptions 1 and 2, a and g must both be positive and therefore the sign of λ depends on whether g is greater than or less than a. If $g = a$ then λ is infinitely large and the model breaks down.

10.6 A CONSUMPTION MODEL

A further example of a simple economic model is provided by the equations

$$C_t = \alpha Y_{t-1} + B \tag{1}$$

$$Y_t \equiv C_t + I_t \tag{2}$$

where
$C_t = $ consumption expenditure

$Y_t = $ national income

$I_t = $ investment expenditure

If I_t is constant in each year—for example, $I_t = F$—then substituting from (2) into (1) gives

$$C_t = \alpha C_{t-1} + (\alpha F + B)$$

which is a first-order non-homogeneous difference equation. The solution of the homogeneous part is

$$C_t = k\alpha^t$$

and the equilibrium solution is

$$C_t = \frac{(\alpha F + B)}{1 - \alpha}$$

so that the complete solution is

$$C_t = k\alpha^t + \frac{(\alpha F + B)}{1 - \alpha}$$

10.7 EXERCISES

1. Solve the following difference equations and sketch the time path of Y_t for $t = 0$ to $t = 5$.

268

(a) $Y_{t+1} = 1.5\ Y_t + 3$ with $Y_0 = 2$

(b) $Y_{t+1} = 0.9\ Y_t + 2$ with $Y_0 = 3$

(c) $Y_{t+1} = Y_t$ with $Y_0 = 6$

(d) $Y_{t+1} = 1.1\ Y_t + 4$ with $Y_0 = 1$

2. A savings club collects £15 from each of its members at the beginning of each year. Assuming that the savings increase at 5 per cent per annum compound interest, express the relationship between the values of the savings at the beginning of year $t+1$ and year t in difference-equation form and hence show that the value of the savings after 20 years is £535.8.

3. Show that the equations

$$\text{Demand:} \qquad q_t = 4 - p_t$$
$$\text{Supply:} \qquad q_t = 1 + 1.5\ p_{t-1}$$

with $p_0 = 1$ lead to an 'exploding' cobweb model.

4. Show that the equations

$$\text{Demand:} \qquad q_t = 4 - p_t$$
$$\text{Supply:} \qquad q_t = 1 + 0.5\ p_{t-1}$$

with $p_0 = 1$ lead to a stable cobweb model. What is the equilibrium price?

5. Show that the equations

$$\text{Demand:} \qquad q_t = 8 - 3\ p_t$$
$$\text{Supply:} \qquad q_t = 3\ p_{t-1}$$

with $p_0 = 2$ lead to an oscillating cobweb model.

6. Solve the Harrod–Domar growth model (see 10.5) when $a = 0.9$, $g = 0.3$ and $Y_0 = 100$.

10.8 SAMUELSON'S MULTIPLIER–ACCELERATOR MODEL

This assumes that

1. Consumer demand in any time period is a function of consumer income in the previous time period.

That is $$C_t = c Y_{t-1}$$

where c is a constant which is less than one and is the marginal propensity to consume.

2. Investment in any time period is a function of the difference in income between the two previous time periods.

$$I_t = b(Y_{t-1} - Y_{t-2})$$

where b is the acceleration coefficient, which is a constant.

3. The national output, or income, Y_t, in any time period is made up of the estimated consumer demand and the estimated investment demand in the same period

$$Y_t = C_t + I_t$$

Based on these assumptions we have the model

$$C_t = c Y_{t-1} \tag{1}$$

$$I_t = b(Y_{t-1} - Y_{t-2}) \tag{2}$$

$$Y_t = C_t + I_t \tag{3}$$

Substituting eqs. (1) and (2) into eq. (3) leads to

$$Y_t = c Y_{t-1} + b(Y_{t-1} - Y_{t-2})$$

$$= (c+b) Y_{t-1} - b Y_{t-2}$$

This is a *second-order linear difference equation* because it relates the value of income in time period t to the value of income in the two immediately preceding time periods. It is possible to determine the value of income in any time period if we know the value of income in any two adjacent periods and the value of the constants b and c.

For example, let us suppose that

$$b = 0.6 \quad c = 0.9 \quad Y_0 = 100 \quad Y_1 = 120$$

Then $$Y_2 = (c+b) Y_1 - b Y_0 = 1.5(120) - 0.6(100)$$

$$= 180 - 60 = 120$$

$$Y_3 = 1.5(120) - 0.6(120) = 180 - 72 = 108$$

The value of income in each successive time period can be calculated by continuing the above procedure. But this is not necessary, because it is possible to obtain a solution to a second-order difference equation and to use this solution to determine Y_t in any time period quite simply as we did in the first-order case.

For convenience the equations and their methods of solution are again discussed under appropriate headings.

10.9 SECOND-ORDER LINEAR DIFFERENCE EQUATIONS

10.9.1. HOMOGENEOUS

These can be written in general as $aY_{t+2}+bY_{t+1}+cY_t=0$ and the solution is of the form $Y_t=\lambda^t$ where λ is a constant whose value is as yet undetermined.

If $Y_t=\lambda^t$, then $Y_{t+1}=\lambda^{t+1}$ and $Y_{t+2}=\lambda^{t+2}$. Substituting these values in the difference equation, we obtain

$$a\lambda^{t+2}+b\lambda^{t+1}+c\lambda^t=0$$

$$\lambda^t(a\lambda^2+b\lambda+c)=0$$

Therefore
$$\lambda_1=\frac{-b+\sqrt{(b^2-4ac)}}{2a}$$

$$\lambda_2=\frac{-b-\sqrt{(b^2-4ac)}}{2a}$$

and the solution is of the form

$$Y_t=k_1\lambda_1^t+k_2\lambda_2^t$$

where k_1 and k_2 are arbitrary constants.

The equation $a\lambda^2+b\lambda+c=0$ is known as the *characteristic equation* and is easily formed by substituting λ^2 for Y_{t+2}, λ for Y_{t+1} and 1 for Y_t in the difference equation. When the roots of the characteristic equation are equal, that is $\lambda_1=\lambda_2=\lambda$, the solution is

$$Y_t=(k_1+k_2t)\lambda^t$$

This contains two arbitrary constants, which is a necessary requirement for it to be the solution to a second-order difference equation (compare differential equations, Section 9.4).

271

When the roots of the characteristic equation are complex, that is $\lambda_1 = a + ib$ and $\lambda_2 = a - ib$, the solution is

$$Y_t = k_1(a + ib)^t + k_2(a - ib)^t$$

The solution is not very useful in this form because it is not possible to analyse the future time path of Y from a combination of real and imaginary numbers. But the solution can be transformed into a more suitable form by making the following substitutions.

Let $\qquad\qquad a = r \cos \theta \qquad$ and $\qquad b = r \sin \theta$

Then $\qquad\qquad a^2 + b^2 = r^2 \cos^2 \theta + r^2 \sin^2 \theta$

$$= r^2(\cos^2 \theta + \sin^2 \theta) = r^2$$

or $\qquad\qquad r = \sqrt{(a^2 + b^2)}$

and $\qquad\qquad \theta = \cos^{-1} \dfrac{a}{r} \qquad$ or $\qquad \sin^{-1} \dfrac{b}{r}$

We already know from the Euler relations (see Section 5.12) that for all values of θ,

$$e^{i\theta} = \cos \theta + i \sin \theta$$

Now $(e^{i\theta})^n = e^{in\theta} = \cos n\theta + i \sin n\theta$.

It can be shown, therefore, that

$$(\cos \theta + i \sin \theta)^n = \cos n\theta + i \sin n\theta$$

This is usually referred to as *De Moivre's theorem* and it can be shown that the identity holds for all values of θ.

Making use of the substitutions $a = r \cos \theta$ and $b = r \sin \theta$

$$(a + ib) = r \cos \theta + ir \sin \theta = r(\cos \theta + i \sin \theta)$$

$\therefore \qquad\qquad (a + ib)^t = [r(\cos \theta + i \sin \theta)]^t$

$$= r^t(\cos \theta + i \sin \theta)^t$$

$$= r^t(\cos t\theta + i \sin t\theta)$$

Similarly, from the properties of complex numbers it can be shown that $\qquad (a - ib)^t = r^t(\cos t\theta - i \sin t\theta)$

272

The solution to the difference equation which has complex roots to the characteristic equation can now be expressed in terms of trigonometrical functions:

$$Y_t = k_1(a+ib)^t + k_2(a-ib)^t$$

becomes $Y_t = k_1[r^t(\cos t\theta + i \sin t\theta)] + k_2[r^t(\cos t\theta - i \sin t\theta)]$

The right-hand side of this equation can be arranged in a more convenient form by collecting together the sine and cosine terms:

$$Y_t = r^t[(k_1+k_2) \cos t\theta + i(k_1-k_2) \sin t\theta]$$
$$= r^t(k_3 \cos t\theta + k_4 \sin t\theta)$$

where $k_3 = k_1 + k_2$, $k_4 = i(k_1 - k_2)$, which are both constants.

This result can be compared to the solution obtained for a second-order differential equation

$$y = e^{ax}(k_3 \cos bx + k_4 \sin bx)$$

It was shown in Chapter 9 that it is possible, and often more useful, to express this in the form $y = Ae^{ax} \cos (bx - \epsilon)$. In a similar way it can be shown, although the mathematics are not discussed here, that the solution to a second-order difference equation can be expressed in the alternative form

$$Y_t = Ar^t \cos (t\theta - \epsilon)$$

where A and ϵ are the two arbitrary constants, whose values are determined in a particular case from two pieces of information which are available about the system.

10.9.2. THE NON-HOMOGENEOUS SECOND-ORDER DIFFERENCE EQUATION

This has the general form

$$aY_{t+2} + bY_{t+1} + cY_t = f(t)$$

It can be shown that the solution to this type of equation is the sum of two parts which must be obtained independently. These are usually called the *transient function* and the *equilibrium*, or *particular solution*.

273

The transient function is simply the solution of the homogeneous part of the equation formed by letting $f(t) = 0$ and this can be obtained by the method of the previous section.

The particular solution is any solution which satisfies the difference equation and its form depends upon the form of $f(t)$.

1. $f(t)$ is a constant. For example, if $a Y_{t+2} + b Y_{t+1} + c Y_t = K$, we try a constant as the particular solution, that is $Y_t = Z$ for all t and it follows that $Y_{t+2} = Y_{t+1} = Y_t = Z$.

Substituting these values in the difference equation, we have

$$a Z + b Z + c Z = K$$

That is $\qquad Z(a + b + c) = K \qquad$ or $\qquad Z = \dfrac{K}{a + b + c}$

It will be found that this value of Z satisfies the difference equation and is therefore a particular solution for all cases except those in which

$$a + b + c = 0$$

In such cases we must try the solution

$$Y_t = Zt$$

so that $\qquad Y_{t+1} = Z(t + 1) = Zt + Z$

$$Y_{t+2} = Z(t + 2) = Zt + 2Z$$

Notice that Zt is the product of Z and t. Substituting these values in the difference equation gives

$$a Z(t + 2) + b Z(t + 1) + c Zt = K$$

or $\qquad a Zt + 2a Z + b Zt + b Z + c Zt = K$

$$Z(a + b + c)t + Z(b + 2a) = K$$

If this equation is to be valid for all values of t, it is necessary that both of the following conditions hold:

$$Z(a + b + c) = 0 \qquad \text{and} \qquad Z(b + 2a) = K$$

The second equation gives a value for Z:

$$Z = \frac{K}{b + 2a} \qquad \text{and hence} \qquad Y_t = Zt = \frac{Kt}{b + 2a}$$

provided that $(b+2a)$ is not equal to zero.

2. $f(t)$ is not a constant

    (a) $f(t)$ is a linear function. For example,

$$a Y_{t+2} + b Y_{t+1} + c Y_t = K_1 + K_2 t$$

The particular solution is of the form

$$Y_t = Z_0 + Z_1 t$$

so that

$$Y_{t+1} = Z_0 + Z_1(t+1)$$

and

$$Y_{t+2} = Z_0 + Z_1(t+2)$$

These values are substituted in the difference equation:

$$a[Z_0 + Z_1(t+2)] + b[Z_0 + Z_1(t+1)] + c[Z_0 + Z_1 t] = K_1 + K_2 t$$

This can be rearranged with the constant terms and the terms in t collected separately:

$$(aZ_0 + 2aZ_1 + bZ_0 + bZ_1 + cZ_0) + (aZ_1 + bZ_1 + cZ_1)t = K_1 + K_2 t$$

that is

$$(a+b+c)Z_0 + (2a+b)Z_1 + (a+b+c)Z_1 t = K_1 + K_2 t$$

If this equation is to be true for all values of t, the following two conditions must hold:

$$(a+b+c)Z_0 + (2a+b)Z_1 = K_1 \tag{1}$$

$$(a+b+c)Z_1 = K_2 \tag{2}$$

From eq. (2)

$$Z_1 = \frac{K_2}{a+b+c}$$

and from eqs. (1) and (2)

$$Z_0 = \frac{K_1 - (2a+b)Z_1}{a+b+c}$$

$$= \frac{K_1}{a+b+c} - \frac{2a+b}{a+b+c}\frac{K_2}{a+b+c}$$

$$= \frac{K_1}{a+b+c} - \frac{(2a+b)K_2}{(a+b+c)^2}$$

275

The particular solution can now be expressed in terms of the known constants a, b, c, K_1 and K_2.

$$Y_t = Z_0 + Z_1 t$$

$$= \frac{K_1}{a+b+c} - \frac{(2a+b)K_2}{(a+b+c)^2} + \frac{K_2 t}{a+b+c}$$

$$= \frac{1}{a+b+c}\left[K_1 - K_2\left(\frac{2a+b}{a+b+c} - t\right)\right]$$

(b) $f(t)$ is of the form K^t. For example,

$$aY_{t+2} + bY_{t+1} + cY_t = K^t$$

The particular solution is of the form

$$Y_t = ZK^t$$

so that

$$Y_{t+1} = ZK^{t+1}$$

and

$$Y_{t+2} = ZK^{t+2}$$

These values are substituted in the difference equations

$$aZK^{t+2} + bZK^{t+1} + cZK^t = K^t$$

$$K^t(aK^2 + bK + c)Z = K^t$$

$$\therefore \qquad Z = \frac{1}{aK^2 + bK + c}$$

and the particular solution is

$$Y_t = ZK^t = \frac{K^t}{aK^2 + bK + c}$$

provided that $aK^2 + bK + c$ is not equal to zero.

If this condition is not satisfied then we must try a particular solution of the form

$$Y_t = ZtK^t$$

If this results in a particular solution with a denominator equal to zero then it is necessary to try the substitution

$$Y_t = Zt^2K^t$$

This procedure can be continued until a solution is obtained.

(c) $f(t)$ is a trigonometrical function. For example,

$$a Y_{t+2} + b Y_{t+1} + c Y_t = K \sin t$$

The particular solution is of the form

$$Y_t = Z_0 \sin t + Z_1 \cos t$$

so that
$$Y_{t+1} = Z_0 \sin (t+1) + Z_1 \cos (t+1)$$

and
$$Y_{t+2} = Z_0 \sin (t+2) + Z_1 \cos (t+2)$$

The values of Z_0 and Z_1 can be found by substituting these values in the difference equation and by making use of the following trigonometrical identities.

$$\sin (A+B) \equiv \sin A \cos B + \cos A \sin B$$

$$\cos (A+B) \equiv \cos A \cos B - \sin A \sin B$$

The exercise, which is rather more lengthy than the previous ones, is left for the reader to attempt.

10.10 A CONSUMPTION–INVESTMENT MODEL

The following example shows how the above methods are applied in the analysis of the time path of national income Y_t given the assumptions that the consumption and investment functions are

$$C_t = 0.5 \ Y_{t-1} + 0.19 \ Y_{t-2} \tag{1}$$

$$I_t = 10 + 50(1.05)^t \tag{2}$$

$$Y_t = C_t + I_t \tag{3}$$

$$Y_0 = 100 \quad \text{and} \quad Y_1 = 120 \tag{4}$$

A second-order difference equation is obtained by substituting (1) and (2) into (3).

$$Y_t = 0.5 \ Y_{t-1} + 0.19 \ Y_{t-2} + 10 + 50(1.05)^t$$

or
$$Y_t - 0.5 \ Y_{t-1} - 0.19 \ Y_{t-2} = 10 + 50(1.05)^t$$

The solution to the reduced form is obtained very easily from the characteristic equation

$$\lambda^2 - 0.5\lambda - 0.19 = 0$$

the roots of which are

$$\lambda = \frac{0.5 \pm \sqrt{[(0.5)^2 + 4 \times 0.19]}}{2} = \frac{0.5 \pm 1}{2}$$

$$\therefore \qquad \lambda_1 = 0.75 \qquad \lambda_2 = -0.25$$

and the transient solution is

$$Y_t = k_1(0.75)^t + k_2(-0.25)^t$$

The particular solution can be found by applying the above methods to the two terms on the right-hand side in turn.

$$Y_t - 0.5 \ Y_{t-1} - 0.19 \ Y_{t-2} = 10$$

Particular solution is

$$Y_t = \left(\frac{1}{1 - 0.5 - 0.19}\right) 10 = \frac{10}{0.31} = 32.26$$

For the equation

$$Y_t - 0.5 \ Y_{t-1} - 0.19 \ Y_{t-2} = 50(1.05)^t$$

the particular solution is

$$Y_t = 50 \left[\frac{1}{K^2 - 0.5K - 0.19}\right](1.05)^t$$

Since $K = 1.05$,

$$Y_t = 50 \left[\frac{1}{0.3875}\right](1.05)^t = 129.03(1.05)^t$$

The solution can then be written as

$$Y_t = k_1(0.75)^t + k_2(-0.25)^t + 32.26 + 129.03(1.05)^t$$

and the value of the constants, k_1 and k_2, found from the initial conditions.

278

$$Y_0 = 100$$

$$\therefore \qquad 100 = k_1 + k_2 + 32.26 + 129.03 \qquad (1)$$

$$Y_1 = 120$$

$$\therefore \qquad 120 = 0.75k_1 - 0.25k_2 + 32.26 + 129.03(1.05) \qquad (2)$$

From eq. (1)

$$k_1 + k_2 = 100 - 161.29 = -61.29$$

and from eq. (2)

$$0.75k_1 - 0.25k_2 = 120 - 32.26 - 135.481 = -47.74$$

These two equations can be solved simultaneously with the result

$$k_1 = -63.06 \qquad \text{and} \qquad k_2 = 1.77$$

The complete solution to the difference equation

$$Y_t - 0.5Y_{t-1} - 0.19Y_{t-2} = 10 + 50(1.05)^t$$

and the initial conditions $Y_0 = 100$, $Y_1 = 120$ is given by

$$Y_t = -63.06(0.75)^t + 1.77(-0.25)^t + 32.26 + 129.03(1.05)^t$$

It is easy to check that this solution satisfies the initial conditions, but not as easy to show that it satisfies the difference equation.

The general shape of the time path of national income can be obtained by considering the function in terms of the four constituent parts. The graphs of these can be very easily sketched as shown in Fig 10.5, where the term $129.03(1.05)^t$ becomes predominant with time.

10.11 EXERCISES

1. Show that Samuelson's Multiplier–Accelerator model leads to a characteristic equation with

    (a) complex roots when $c = 0.9$, $b = 0.5$

    (b) repeated roots when $c = 0.96$, $b = 0.64$

    (c) real roots when $c = 0.9$, $b = 0.3$

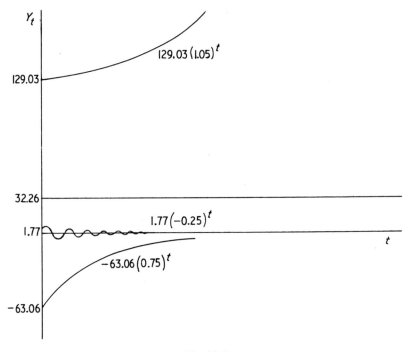

Fig 10.5

2. Solve the following difference equations and sketch the time path of Y_t for $t=0$ to $t=5$ for (a)–(e).

(a) $Y_{t+2} - Y_{t+1} - 2Y_t = 3$ with $Y_0 = 2$ $Y_1 = 2$

(b) $Y_{t+2} - 3Y_{t+1} + 2Y_t = 4$ with $Y_0 = 1$ $Y_1 = 2$

(c) $Y_{t+2} - 4Y_{t+1} + 4Y_t = 3^t$ with $Y_0 = 3$ $Y_1 = 4$

(d) $Y_{t+2} - 3Y_{t+1} + 2Y_t = 3(2^t)$ with $Y_0 = 1$ $Y_1 = 5$

(e) $4Y_{t+2} + 4Y_{t+1} + Y_t = 2 + 3t$ with $Y_1 = 1$ $Y_2 = 2$

(f) $Y_{t+2} + 4Y_{t+1} + 5Y_t = 4$

(g) $Y_{t+2} - 4Y_{t+1} + 8Y_t = t$

3. For the equations

$$Y_t = C_t + I_t \tag{1}$$

$$C_t = \alpha Y_t + \beta Y_{t-1} + \delta \tag{2}$$

$$I_t = \gamma(Y_{t-1} - Y_{t-2}) \tag{3}$$

show that substitution from (2) and (3) into (1) produces a second-order difference equation which, if $\delta = 0$, is similar to the equation from Samuelson's multiplier–accelerator model (Section 10.8).

Solve the equation when

(a) $\alpha = 0.6$, $\beta = 0.2$, $\gamma = 0.3$ and $\delta = 1$

(b) $\alpha = 0.5$, $\beta = 0.2$, $\gamma = 0.2$ and $\delta = 1$

APPENDIX A

Trigonometric Functions

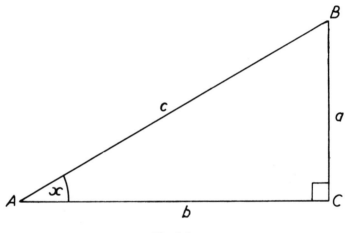

Fig A.1

A.1 DEFINITIONS

Let us consider a right-angled triangle ABC (Fig A.1). If the side BC is of length a, AC of length b, AB of length c, and x is the angle BAC then we define

$$\text{sine } x = \sin x = \frac{a}{c} \tag{1}$$

$$\text{cosine } x = \cos x = \frac{b}{c} \tag{2}$$

$$\text{tangent } x = \tan x = \frac{a}{b} \tag{3}$$

Values of these functions can be obtained from tables.

For a right-angled triangle,

$$a^2 + b^2 = c^2$$

or, dividing by c^2,

$$\left(\frac{a}{c}\right)^2 + \left(\frac{b}{c}\right)^2 = 1$$

That is

$$(\sin x)^2 + (\cos x)^2 = 1$$

or

$$\sin^2 x + \cos^2 x = 1 \tag{4}$$

Therefore,

$$\sin^2 x = 1 - \cos^2 x$$

$$\cos^2 x = 1 - \sin^2 x$$

Notice that

$$\tan x = \frac{a}{b} = \frac{a/c}{b/c} = \frac{\sin x}{\cos x}$$

From (4), by dividing by $\cos^2 x$,

$$\frac{\sin^2 x}{\cos^2 x} + \frac{\cos^2 x}{\cos^2 x} = \frac{1}{\cos^2 x}$$

or,

$$\tan^2 x + 1 = \sec^2 x \tag{5}$$

where

$$\sec x = \text{secant } x = 1/\cos x$$

Two other trigonometric functions are the cosecant and cotangent defined by

$$\text{cosecant } x = \text{cosec } x = \frac{1}{\sin x}$$

and

$$\text{cotangent } x = \text{cotan } x = \frac{1}{\tan x} = \frac{\cos x}{\sin x}$$

A.2 COMPOUND ANGLES

For any two angles x and y the following relationships can be shown to be true:

$$\sin(x + y) = \sin x \cos y + \cos x \sin y \tag{6}$$

$$\cos(x + y) = \cos x \cos y - \sin x \sin y \tag{7}$$

283

$$\sin (x-y) = \sin x \cos y - \cos x \sin y \qquad (8)$$

$$\cos (x-y) = \cos x \cos y + \sin x \sin y \qquad (9)$$

$$\tan (x+y) = \frac{\tan x + \tan y}{1 - \tan x \tan y} \qquad (10)$$

$$\tan (x-y) = \frac{\tan x - \tan y}{1 + \tan x \tan y} \qquad (11)$$

Using these relationships, we have

(a) when $x=y$, from (6)

$$\sin (x+x) = \sin 2x = 2 \sin x \cos x \qquad (12)$$

(b) when $x=y$, from (7),

$$\cos 2x = \cos x \cos x - \sin x \sin x$$

$$= \cos^2 x - \sin^2 x$$

$$= 1 - 2 \sin^2 x = 2 \cos^2 x - 1 \qquad (13)$$

since $\cos^2 x + \sin^2 x = 1$. Hence,

$$2 \sin^2 x = 1 - \cos 2x$$

or $\qquad\qquad\qquad\qquad \sin^2 x = \dfrac{1 - \cos 2x}{2} \qquad (14)$

and $\qquad\qquad\qquad\qquad \cos^2 x = \dfrac{1 + \cos 2x}{2} \qquad (15)$

(c) When $x=y$, from (10)

$$\tan 2x = \frac{2 \tan x}{1 - \tan^2 x} \qquad (16)$$

$$\sin x + \sin y = 2 \sin \left(\frac{x+y}{2}\right) \cos \left(\frac{x-y}{2}\right) \qquad (17)$$

$$\sin x - \sin y = 2 \cos \left(\frac{x+y}{2}\right) \sin \left(\frac{x-y}{2}\right) \qquad (18)$$

284

$$\cos x + \cos y = 2 \cos \left(\frac{x+y}{2}\right) \cos \left(\frac{x-y}{2}\right) \qquad (19)$$

$$\cos x - \cos y = -2 \sin \left(\frac{x+y}{2}\right) \sin \left(\frac{x-y}{2}\right) \qquad (20)$$

The formulae (17) to (20) can be verified by using (6) to (9) to expand the right-hand side of each statement.

A.3 DEGREE AND RADIANS

In elementary mathematical work it is convenient to measure angles in degrees. In more advanced work, however, it is more convenient to use another measure of angles, known as radians. A *radian* is defined as the angle at the centre of a circle which contains an arc length equal to the radius of the circle (see Fig A.2).

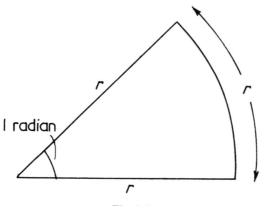

Fig A.2

If r is the radius of a circle, the circumference is of length $2\pi r$. The angle at the centre of a circle is $360°$, and so

$$360° = \frac{2\pi r}{r} = 2\pi \text{ radians}$$

or
$$\pi \text{ radians} = 180°$$

and
$$\pi/2 \text{ radians} = 90°$$

285

A.4   GENERAL ANGLES

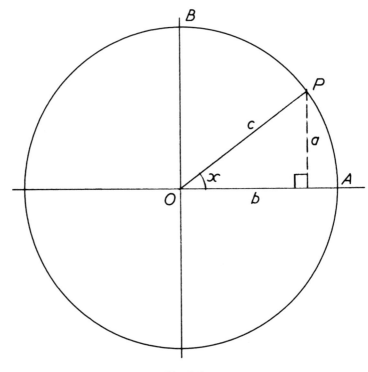

Fig A.3

From Fig A.3 and the definitions $\sin x = a/c$, $\cos x = b/c$, $\tan x = a/b$ we can see that varying x causes the point p to mark out a quadrant of a circle from A to B. At the same time the values of a and b vary.

If we allow x to become very small and approach zero, a approaches zero, and b approaches c, hence $\sin 0 = 0$, $\cos 0 = 1$, $\tan 0 = 0$.

Similarly if we allow x to approach $90°$ a approaches c and b approaches zero. Hence,

$$\sin 90° = \sin (\pi/2) = 1$$
$$\cos 90° = \cos (\pi/2) = 0$$
$$\tan 90° = \tan (\pi/2) = \infty$$

Let us now increase x to $135°$ or $3\pi/4$ radians (Fig A.4). We know that, for example, using (6),

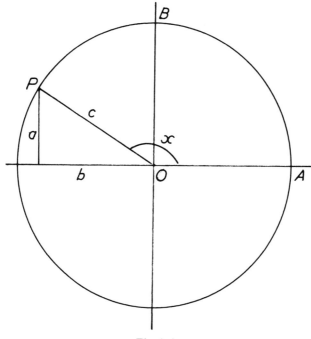

Fig A.4

$$\sin 135° = \sin (90 + 45)° = \sin 90° \cos 45° + \cos 90° \sin 45°$$
$$= \cos 45° = 0.7071,$$

that is, we can obtain $\sin x$ even when x is greater than $90°$ or $\pi/2$ radians. Similarly it can be shown that

$$\cos 135° = -\sin 45° = -0.7071$$
$$\tan 135° = -\tan 45° = -1.0000$$

Notice that $\cos 135°$ and $\tan 135°$ are both negative. This is because

$$\cos x = \frac{b}{c} \qquad \tan x = \frac{a}{b}$$

287

and b is measured in the negative direction from 0, and so is a negative number. Both a and c are positive numbers.

Fig A.5 shows the angle $x = 225°$. In this case both a and b are negative numbers so that sin x and cos x are negative, but tan x is positive. It can be shown that

$$\sin 225° = -0.7071 \qquad \cos 225° = -0.7071 \qquad \tan 225° = 1.000$$

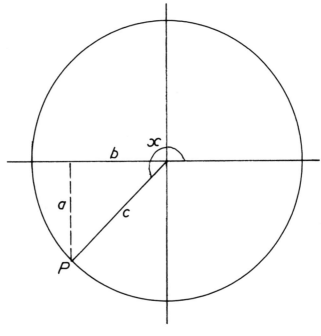

Fig A.5

Similarly, the angle x can be increased to, say, 315°, and in this case a is negative and both b and c are positive, so that sin x and tan x are negative and cos x is positive. These results can be summarised as in Fig A.6 to indicate which of sine, cosine and tangent are positive.

So far we have considered only angles of up to 2π radians (360°), but it is possible to allow an angle to exceed this by rotating the point

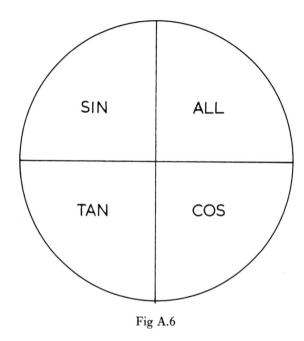

Fig A.6

P (Fig A.5) around the circle. In this way, an angle of, say, $2\pi + x$ has the same sine, cosine, and tangent as x has.

We know that

$$\sin 0 = 0 \qquad \cos 0 = 1 \qquad \tan 0 = 0$$
$$\sin (\pi/2) = 1 \quad \cos (\pi/2) = 0 \quad \tan (\pi/2) = \infty$$

By evaluating $\sin x$, $\cos x$, and $\tan x$ for other values of x we can draw the graphs of these functions (Fig A.7).

The graph of $\cos x$ is identical to the graph of $\sin x$ moved $\pi/2$ to the right. The graph of $\tan x$ has asymptotes at $\pi/2$, $3\pi/2$, $5\pi/2$, $\ldots$ and so it consists of a series of sections.

A.5 DIFFERENTIATION

If $y = \sin x$ and we allow x to increase by a small amount δx, and let the corresponding increase in y be δy, then

$$y + \delta y = \sin (x + \delta x)$$

289

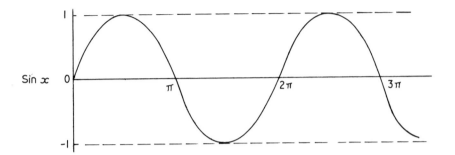

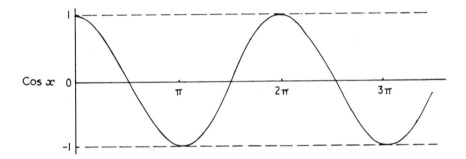

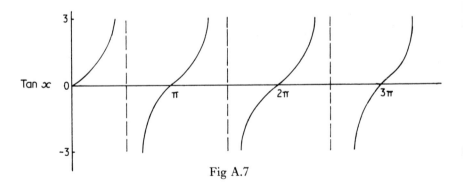

Fig A.7

Subtracting, we have

$$\delta y = \sin(x + \delta x) - \sin x$$

Using (6), we obtain

$$\delta y = \sin x \cos \delta x + \cos x \sin \delta x - \sin x$$

$$= \sin x (\cos \delta x - 1) + \cos x \sin \delta x$$

Hence,
$$\frac{\delta y}{\delta x} = \sin x \left(\frac{\cos \delta x - 1}{\delta x} \right) + \frac{\cos x \sin \delta x}{\delta x}$$

Now by Maclaurin's theorem (Section 5.11)

$$\cos x = 1 - \frac{x^2}{2!} + \frac{x^4}{4!} - \frac{x^6}{6!} + \cdots$$

and
$$\sin x = x - \frac{x^3}{3!} + \frac{x^5}{5!} - \frac{x^7}{7!} + \cdots$$

That is,
$$\cos \delta x = 1 - \frac{(\delta x)^2}{2!} + \frac{(\delta x)^4}{4!} - \frac{(\delta x)^6}{6!} + \cdots$$

so that
$$\frac{\cos \delta x - 1}{\delta x} = - \frac{\delta x}{2!} + \frac{(\delta x)^3}{4!} - \frac{(\delta x)^5}{6!} + \cdots$$

and
$$\lim_{\delta x \to 0} \left(\frac{\cos \delta x - 1}{\delta x} \right) = 0$$

Similarly,
$$\frac{\sin \delta x}{\delta x} = 1 - \frac{(\delta x)^2}{3!} + \frac{(\delta x)^4}{5!} - \frac{(\delta x)^6}{7!} + \cdots$$

and
$$\lim_{\delta x \to 0} \left(\frac{\sin \delta x}{\delta x} \right) = 1$$

Therefore,
$$\lim_{\delta x \to 0} \left(\frac{\delta y}{\delta x} \right) = \frac{dy}{dx} = \sin x \, (0) + \cos x \, (1) = \cos x$$

Hence, if

$$y = \sin x \qquad \frac{dy}{dx} = \cos x$$

The same method can be used to show that if $y = \cos x$, $dy/dx = -\sin x$.

If $y = \tan x = (\sin x)/(\cos x)$, then, using the rule for differentiating quotients, we have

$$\frac{dy}{dx} = \frac{\cos x\,(\cos x) - \sin x\,(-\sin x)}{(\cos x)^2}$$

$$= \frac{\cos^2 x + \sin^2 x}{\cos^2 x} = \frac{1}{\cos^2 x} = \sec^2 x$$

Similarly it can be shown that if

$$y = \cotan x = \frac{1}{\tan x} \qquad \frac{dy}{dx} = -\cosec^2 x = \frac{-1}{\sin^2 x}$$

$$y = \cosec x = \frac{1}{\sin x} \qquad \frac{dy}{dx} = \frac{-\cos x}{\sin^2 x} = \frac{-\cosec x}{\tan x}$$

$$y = \sec x = \frac{1}{\cos x} \qquad \frac{dy}{dx} = \frac{\sin x}{\cos^2 x} = \tan x \sec x$$

The above results can be generalised for the angle mx (where m is a constant) by use of the function of a function rule.

For example, if $y = \sin mx$, let $\theta = mx$ so that $d\theta/dx = m$.

Then $\qquad\qquad\qquad y = \sin \theta \qquad \dfrac{dy}{d\theta} = \cos \theta$

$\therefore \qquad\qquad\qquad \dfrac{dy}{dx} = \dfrac{dy}{d\theta}\dfrac{d\theta}{dx} = m \cos mx$

The derivatives of the trigonometric functions are listed in Table A.1.

A.6   INTEGRATION

The integrals of some trigonometric functions can be deduced from Table A.1. For example,

$$\int \cos x\,dx = \sin x + c$$

and $\qquad\qquad\qquad \int \sin x\,dx = -\cos x + c$

Now $\qquad\qquad \int \tan x\,dx = \int \dfrac{\sin x}{\cos x}\,dx = -\int \dfrac{-\sin x}{\cos x}\,dx$

$$= -\log(\cos x) + c$$

TABLE A.1

Function	Derivative	Function	Derivative
$\sin x$	$\cos x$	$\sin mx$	$m \cos mx$
$\cos x$	$-\sin x$	$\cos mx$	$-m \sin mx$
$\tan x$	$\sec^2 x$	$\tan mx$	$m \sec^2 mx$
$\cotan x$	$-\cosec^2 x$	$\cotan mx$	$-m \cosec^2 mx$
$\sec x$	$\tan x \sec x$	$\sec mx$	$m \tan mx \sec mx$
$\cosec x$	$\dfrac{-\cosec x}{\tan x}$	$\cosec mx$	$\dfrac{-m \cosec mx}{\tan mx}$

The integrals of other trigonometric functions are more difficult but some can be obtained by making the substitutions

$$\sin x = \frac{2 \tan (x/2)}{1 + \tan^2 (x/2)} \qquad \cos x = \frac{1 - \tan^2 (x/2)}{1 + \tan^2 (x/2)}$$

or, writing $t = \tan (x/2)$,

$$\sin x = \frac{2t}{1 + t^2} \qquad \cos x = \frac{1 - t^2}{1 + t^2}$$

These can be shown to be true by using the formulae for compound angles:

$$\frac{2 \tan (x/2)}{1 + \tan^2 (x/2)} = \frac{2[\sin (x/2)]/[\cos (x/2)]}{1 + [\sin^2 (x/2)]/[\cos^2 (x/2)]}$$

$$= \frac{2 \sin (x/2)}{\cos (x/2)[\cos^2 (x/2) + \sin^2 (x/2)]/\cos^2 (x/2)}$$

$$= \frac{2 \sin (x/2) \cos (x/2)}{1} = \sin x, \text{ using (12) and (4)}$$

293

Similarly,

$$\frac{1 - \tan^2 (x/2)}{1 + \tan^2 (x/2)} = \frac{1 - [\sin^2 (x/2)]/[\cos^2 (x/2)]}{1 + [\sin^2 (x/2)]/[\cos^2 (x/2)]}$$

$$= \frac{\cos^2 (x/2) - \sin^2 (x/2)}{\cos^2 (x/2) + \sin^2 (x/2)} = \cos x$$

using (4) and (13). These can be used to integrate some functions which cannot be integrated by other methods.

For example
$$\int \frac{dx}{\cos x}$$

Let $t = \tan (x/2)$

Then
$$\cos x = \frac{1 - t^2}{1 + t^2} \qquad \sin x = \frac{2t}{1 + t^2}$$

$$dx = \frac{2 \, dt}{1 + t^2}$$

Now
$$\int \frac{dx}{\cos x} = \int \frac{(1 + t^2)}{(1 - t^2)} \frac{2 \, dt}{(1 + t^2)} = \int \frac{2 \, dt}{1 - t^2}$$

$$= \int \frac{1}{1 + t} \, dt + \frac{1}{1 - t} \, dt$$

$$= \log (1 + t) - \log (1 - t) + C$$

$$= \log \left(\frac{1 + t}{1 - t} \right) + C$$

$$= \log \left(\frac{1 + \tan (x/2)}{1 - \tan (x/2)} \right) + C$$

Some standard integrals are given in Table A.2 (the constants of integration are omitted).

A.7 INVERSE FUNCTIONS

If $y = \sin x$ then $x = \sin^{-1} y$ is known as the *inverse sine function*, or, x is an angle whose sine is y. Similarly, $x = \cos^{-1} y$ and $x = \tan^{-1} y$ are the *inverse cosine* and *tangent functions*.

TABLE A.2

Function	Integral	Function	Integral
$\sin x$	$-\cos x$	$\sin ax$	$-\dfrac{1}{a}\cos ax$
$\cos x$	$\sin x$	$\cos ax$	$\dfrac{1}{a}\sin ax$
$\tan x$	$-\log(\cos x)$	$\tan ax$	$-\dfrac{1}{a}\log(\cos ax)$
$\cotan x$	$\log(\sin x)$	$\cotan ax$	$\dfrac{1}{a}\log(\sin ax)$

These are useful in integration when trigonometric substitutions can be made. For example,

$$\int \frac{1}{\sqrt{(a^2-x^2)}}\,dx$$

Substitute
$$x = a\sin\theta$$
$$dx = a\cos\theta\,d\theta$$

so that
$$\int \frac{1}{\sqrt{(a^2-x^2)}}\,dx = \int \frac{a\cos\theta\,d\theta}{\sqrt{(a^2-a^2\sin^2\theta)}} = \int \frac{a\cos\theta\,d\theta}{a\cos\theta}$$
$$= \int 1\,d\theta = \theta + C$$

but if $x = a\sin\theta$, $\theta = \sin^{-1} x/a$, hence

$$\int \frac{dx}{\sqrt{(a^2-x^2)}} = \sin^{-1}(x/a) + C$$

In the same way,

$$\int \frac{dx}{\sqrt{(a^2+x^2)}} = \frac{1}{a}\tan^{-1}\left(\frac{x}{a}\right) + C$$

where the appropriate substitution is $x = a\tan\theta$.

295

APPENDIX B

Answers to Exercises

EXERCISES 1.5
1. (a) $Y = 3x + 4$; intercept is 4, slope is 3
   (b) $Y = 3x - 4$; intercept is -4, slope is 3
   (c) $Y = 4 - 3x$; intercept is 4, slope is -3
   (d) $Y = 3x$;   intercept is 0, slope is 3
   (e) $Y = 4$;   intercept is 4, slope is 0.

2. Let $TC = a + bq$
   then from (a) $70 = a + 10b$
   and (b) $120 = a + 20b$

   Solving these equations simultaneously gives the values $a = 20$, $b = 5$

   $\therefore$ $TC = 20 + 5q$

   Fixed cost $= 20$, variable cost $= 5$.

3. $TC = 15 + 6q$
   Total cost is £75 for an output of 10 units.

4. (b) Fixed cost is 3, variable cost is 2   (c) 33   (d) 21.

EXERCISE   1.8
1. $Y = 33 + 2x$
   (a) at the breakeven point $Y =$ total revenue $= 13x$
   $\therefore$ $13x = 33 + 2x$
   $\therefore$ $x = 3$
   (b) Net revenue $= pX - Y = 13(15) - 33 - 2(15) = 132$
   (d) because the slope is 2, ie the marginal cost is £2 per unit.

296

2. (a) $TC = 50 + 5q$
   (b) At breakeven point, $50 + 5q = 10q$ so that $q = 10$
   (c) Net revenue $= 12p - 110$

   (i) $p = 5$, net revenue $= -£50$
   (ii) $p = 10$, net revenue $= £10$
   (iii) $p = 15$, net revenue $= £70$

3. (a) At equilibrium $q_s = q_d$
   $\therefore \quad 25p - 10 = 200 - 5p$
   $\therefore \quad p = 7 \quad$ and $\quad q = 165$
   (b) $20p - 25 = 200 - 5p$
   $\therefore \quad p = 9 \quad$ and $\quad q = 155$

4. (a) $p = 4, \quad q = 3$
   (b) $p = \frac{1}{2}, \quad q = 0$
   (c) there is no unique solution
   (d) the equations are inconsistent.

EXERCISES 1.10

1. $2x_1 + 2x_2 = 2$

   $3x_1 - x_2 = 1$

   $$\therefore \quad x_1 = \frac{\begin{vmatrix} 2 & 2 \\ 1 & -1 \end{vmatrix}}{\begin{vmatrix} 2 & 2 \\ 3 & -1 \end{vmatrix}} = \frac{(-2 \quad -2)}{(-2 \quad -6)} = \frac{-4}{-8} = \frac{1}{2}$$

   and
   $$x_2 = \frac{\begin{vmatrix} 2 & 2 \\ 3 & 1 \end{vmatrix}}{\begin{vmatrix} 2 & 2 \\ 3 & -1 \end{vmatrix}} = \frac{2 - 6}{-8} = \frac{-4}{-8} = \frac{1}{2}$$

2. $x_1 = 1, \ x_2 = 1$.

3. $x_1 - x_2 + x_3 = 4$

   $x_1 + x_2 + 3x_3 = 8$

   $x_1 + 2x_2 - x_3 = 0$

$$\therefore x_1 = \frac{\begin{vmatrix} 4 & -1 & 1 \\ 8 & 1 & 3 \\ 0 & 2 & -1 \end{vmatrix}}{\begin{vmatrix} 1 & -1 & 1 \\ 1 & 1 & 3 \\ 1 & 2 & -1 \end{vmatrix}} = \frac{4\begin{vmatrix} 1 & 3 \\ 2 & -1 \end{vmatrix} -(-1)\begin{vmatrix} 8 & 3 \\ 0 & -1 \end{vmatrix} +1\begin{vmatrix} 8 & 1 \\ 0 & 2 \end{vmatrix}}{1\begin{vmatrix} 1 & 3 \\ 2 & -1 \end{vmatrix} -(-1)\begin{vmatrix} 1 & 3 \\ 1 & -1 \end{vmatrix} +1\begin{vmatrix} 1 & 1 \\ 1 & 2 \end{vmatrix}}$$

$$= \frac{4(-7)+(-8)+(16)}{(-7)+(-4)+(1)} = \frac{-20}{-10} = 2$$

$$x_2 = \frac{\begin{vmatrix} 1 & 4 & 1 \\ 1 & 8 & 3 \\ 1 & 0 & -1 \end{vmatrix}}{\begin{vmatrix} 1 & -1 & 1 \\ 1 & 1 & 3 \\ 1 & 2 & -1 \end{vmatrix}} = \frac{-8+16-8}{-10} = 0$$

$$x_3 = \frac{\begin{vmatrix} 1 & -1 & 4 \\ 1 & 1 & 8 \\ 1 & 2 & 0 \end{vmatrix}}{\begin{vmatrix} 1 & -1 & 1 \\ 1 & 1 & 3 \\ 1 & 2 & -1 \end{vmatrix}} = \frac{-16-8+4}{-10} = \frac{-20}{-10} = 2$$

4. $x_1 = 1, \quad x_2 = 1, \quad x_3 = -1$

5. $x_1 = 2, \quad x_2 = 2, \quad x_3 = 2$

6. $x_1 = -1, \quad x_2 = -2, \quad x_3 = -3$

EXERCISES 2.3

1.
$$\mathbf{A} = \begin{bmatrix} 2 & 4 \\ 1 & 3 \end{bmatrix} \qquad \mathbf{B} = \begin{bmatrix} 2 & 0 \\ -1 & 1 \end{bmatrix} \qquad \mathbf{C} = \begin{bmatrix} 3 \\ 1 \end{bmatrix}$$

(a) $\mathbf{A} + \mathbf{B} = \begin{bmatrix} 2+2 & 4+0 \\ 1+(-1) & 3+1 \end{bmatrix} = \begin{bmatrix} 4 & 4 \\ 0 & 4 \end{bmatrix}$

(b) $\mathbf{A} - 2\mathbf{B} = \begin{bmatrix} 2-(2 \times 2) & 4-(2 \times 0) \\ 1-(2 \times -1) & 3-(2 \times 1) \end{bmatrix} = \begin{bmatrix} -2 & 4 \\ 3 & 1 \end{bmatrix}$

(c) $\mathbf{AB} = \begin{bmatrix} 2 & 4 \\ 1 & 3 \end{bmatrix} \begin{bmatrix} 2 & 0 \\ -1 & 1 \end{bmatrix} = \begin{bmatrix} (2 \times 2)+(4 \times -1) & (2 \times 0)+(4 \times 1) \\ (1 \times 2)+(3 \times -1) & (1 \times 0)+(3 \times 1) \end{bmatrix}$

$= \begin{bmatrix} 4-4 & 0+4 \\ 2-3 & 0+3 \end{bmatrix} = \begin{bmatrix} 0 & 4 \\ -1 & 3 \end{bmatrix}$

(d) $\mathbf{AC} = \begin{bmatrix} 2 & 4 \\ 1 & 3 \end{bmatrix} \begin{bmatrix} 3 \\ 1 \end{bmatrix} = \begin{bmatrix} (2 \times 3)+(4 \times 1) \\ (1 \times 3)+(3 \times 1) \end{bmatrix} = \begin{bmatrix} 10 \\ 6 \end{bmatrix}$

$\qquad\quad (2 \times 2) \ (2 \times 1) \qquad\qquad\qquad\qquad (2 \times 1)$

(e) $\mathbf{CA} = \begin{bmatrix} 3 \\ 1 \end{bmatrix} \begin{bmatrix} 2 & 4 \\ 1 & 3 \end{bmatrix}$ It is not possible to form the product matrix $\mathbf{CA}$

$\qquad\quad (2 \times 1) \ (2 \times 2)$

(f) $\mathbf{B} = \begin{bmatrix} 2 & 0 \\ -1 & 1 \end{bmatrix} \qquad \mathbf{B}' = \begin{bmatrix} 2 & -1 \\ 0 & 1 \end{bmatrix}$

$\mathbf{AB}' = \begin{bmatrix} 2 & 4 \\ 1 & 3 \end{bmatrix} \begin{bmatrix} 2 & -1 \\ 0 & 1 \end{bmatrix}$

$= \begin{bmatrix} (2 \times 2)+(4 \times 0) & (2 \times -1)+(4 \times 1) \\ (1 \times 2)+(3 \times 0) & (1 \times -1)+(3 \times 1) \end{bmatrix}$

$= \begin{bmatrix} 4 & 2 \\ 2 & 2 \end{bmatrix}$

(g) $\mathbf{C} = \begin{bmatrix} 3 \\ 1 \end{bmatrix} \qquad \mathbf{C}' = [3 \quad 1]$

$$\mathbf{C'A} = [3 \quad 1] \begin{bmatrix} 2 & 4 \\ 1 & 3 \end{bmatrix}$$

$$(1 \times 2) \quad (2 \times 2)$$

$$= [(3 \times 2) + (1 \times 1) \quad (3 \times 4) + (1 \times 3)] = [7 \quad 15]$$
$$(1 \times 2)$$

(h) $\mathbf{BA} = \begin{bmatrix} 2 & 0 \\ -1 & 1 \end{bmatrix} \begin{bmatrix} 2 & 4 \\ 1 & 3 \end{bmatrix}$

$$= \begin{bmatrix} (2 \times 2) + (0 \times 1) & (2 \times 4) + (0 \times 3) \\ (-1 \times 2) + (1 \times 1) & (-1 \times 4) + (1 \times 3) \end{bmatrix} = \begin{bmatrix} 4 & 8 \\ -1 & -1 \end{bmatrix}$$

This is not equal to $\mathbf{AB}$ (obtained earlier in (c)).

2.
$$\mathbf{A} = \begin{bmatrix} 1 & 0 & 2 \\ 1 & 1 & 3 \\ 0 & 2 & -1 \end{bmatrix} \quad \mathbf{B} = \begin{bmatrix} 2 & 1 \\ 1 & -1 \\ 2 & 2 \end{bmatrix} \quad \mathbf{C} = [1 \quad 0 \quad 2]$$

(a) $\mathbf{AB} = \begin{bmatrix} 1 & 0 & 2 \\ 1 & 1 & 3 \\ 0 & 2 & -1 \end{bmatrix} \begin{bmatrix} 2 & 1 \\ 1 & -1 \\ 2 & 2 \end{bmatrix}$

$$= \begin{bmatrix} (1 \times 2) + (0 \times 1) + (2 \times 2) & (1 \times 1) + (0 \times -1) + (2 \times 2) \\ (1 \times 2) + (1 \times 1) + (3 \times 2) & (1 \times 1) + (1 \times -1) + (3 \times 2) \\ (0 \times 2) + (2 \times 1) + (-1 \times 2) & (0 \times 1) + (2 \times -1) + (-1 \times 2) \end{bmatrix}$$

$$= \begin{bmatrix} 6 & 5 \\ 9 & 6 \\ 0 & -4 \end{bmatrix}$$

(b) $\mathbf{A'B} = \begin{bmatrix} 1 & 1 & 0 \\ 0 & 1 & 2 \\ 2 & 3 & -1 \end{bmatrix} \begin{bmatrix} 2 & 1 \\ 1 & -1 \\ 2 & 2 \end{bmatrix}$

$$(3 \times 3) \qquad (3 \times 2)$$

$$= \begin{bmatrix} (1 \times 2) + (1 \times 1) + (0 \times 2) & (1 \times 1) + (1 \times -1) + (0 \times 2) \\ (0 \times 2) + (1 \times 1) + (2 \times 2) & (0 \times 1) + (1 \times -1) + (2 \times 2) \\ (2 \times 2) + (3 \times 1) + (-1 \times 2) & (2 \times 1) + (3 \times -1) + (-1 \times 2) \end{bmatrix}$$

$$= \begin{bmatrix} 3 & 0 \\ 5 & 3 \\ 5 & -3 \end{bmatrix}$$

(3×3)

(c)   **A**   **C**   does not exist

   (3×3) (1×3)

(d) $\mathbf{AC'} = \begin{bmatrix} 1 & 0 & 2 \\ 1 & 1 & 3 \\ 0 & 2 & -1 \end{bmatrix} \begin{bmatrix} 1 \\ 0 \\ 2 \end{bmatrix}$

$$= \begin{bmatrix} (1 \times 1) + (0 \times 0) + (2 \times 2) \\ (1 \times 1) + (1 \times 0) + (3 \times 2) \\ (0 \times 1) + (2 \times 0) + (-1 \times 2) \end{bmatrix}$$

$$= \begin{bmatrix} 5 & 7 & -2 \end{bmatrix}$$

(e) $\mathbf{CB} = \begin{bmatrix} 1 & 0 & 2 \end{bmatrix} \begin{bmatrix} 2 & 1 \\ 1 & -1 \\ 2 & 2 \end{bmatrix}$

     (1×3) (3×2)

$$= \begin{bmatrix} (1 \times 2) + (0 \times 1) + (2 \times 2) & (1 \times 1) + (0 \times -1) + (2 \times 2) \end{bmatrix}$$

$$= \begin{bmatrix} 6 & 5 \end{bmatrix}$$

   (1×2)

EXERCISES 2.5

1. $\mathbf{A} = \begin{bmatrix} 2 & 1 \\ 1 & 3 \end{bmatrix}$

$\therefore \;\; \mathbf{C} = \begin{bmatrix} 3 & -1 \\ -1 & 2 \end{bmatrix}$ and $|\mathbf{A}| = 5$

$\therefore \;\; \mathbf{A}^{-1} = \frac{1}{5} \begin{bmatrix} 3 & -1 \\ -1 & 2 \end{bmatrix} = \begin{bmatrix} \frac{3}{5} & -\frac{1}{5} \\ -\frac{1}{5} & \frac{2}{5} \end{bmatrix}$

MATHEMATICS FOR ECONOMISTS

The solution to the equations is given by

$$\mathbf{X} = \mathbf{A}^{-1}\mathbf{b} = \begin{bmatrix} \frac{3}{5} & -\frac{1}{5} \\ -\frac{1}{5} & \frac{2}{5} \end{bmatrix} \begin{bmatrix} 4 \\ 7 \end{bmatrix}$$

$$= \begin{bmatrix} (\frac{3}{5} \times 4) + (-\frac{1}{5} \times 7) \\ (-\frac{1}{5} \times 4) + (\frac{2}{5} \times 7) \end{bmatrix} = \begin{bmatrix} \frac{12}{5} - \frac{7}{5} \\ -\frac{4}{5} + \frac{14}{5} \end{bmatrix} = \begin{bmatrix} 1 \\ 2 \end{bmatrix}$$

$$\therefore \quad x_1 = 1, \qquad x_2 = 2$$

2. (a) $x_1 = 58/23$, $\quad x_2 = -14/23$
   (b) $x_1 = 0$, $\quad x_2 = -1$, $\quad x_3 = 3$
   (c) $x_1 = 2$, $\quad x_2 = 0$, $\quad x_3 = 1$

EXERCISES 2.7

1. (a)
$$\begin{bmatrix} 1 & 1 & 3 \\ 2 & 1 & 1 \\ 1 & -2 & -1 \end{bmatrix}$$

Subtract twice row one from row two, and row one from row three.

$$\begin{bmatrix} 1 & 1 & 3 \\ 0 & -1 & -5 \\ 0 & -3 & -4 \end{bmatrix}$$

add row two to row one, subtract three times row two from row three, and multiply row two by minus one.

$$\begin{bmatrix} 1 & 0 & -2 \\ 0 & 1 & 5 \\ 0 & 0 & 11 \end{bmatrix}$$

divide row three by eleven and then add twice the new row three to row one and subtract five times the new row three from row two

$$\begin{bmatrix} 1 & 0 & 0 \\ 0 & 1 & 0 \\ 0 & 0 & 1 \end{bmatrix}$$

302

this is the unit matrix and therefore the original matrix is of rank three.

(b) rank two

(c) rank three

(d) rank three

2. The determinant $|\mathbf{A}|$ of the coefficients is not equal to zero. Therefore there is a unique set of values of the variables x_1, x_2, x_3 and x_4 which satisfies the equations.

EXERCISES 2.10

1. Level of final demand which can be met by Industry 1 is

$1,500 - (200 + 300) = 1,000$, and by Industry 2 is

$$2,500 - (500 + 100) = 1,900.$$

2. The matrix of technological coefficients is

$$\begin{bmatrix} \frac{200}{1500} & \frac{300}{2500} \\ \frac{500}{1500} & \frac{100}{2500} \end{bmatrix} = \begin{bmatrix} \frac{2}{15} & \frac{3}{25} \\ \frac{1}{3} & \frac{1}{25} \end{bmatrix}$$

and the new situation is as shown in the table.

	Input to		Level of output
	Industry 1	Industry 2	
Industry 1	$\frac{2}{15} \times 2,000$	$\frac{3}{25} \times 2,500$	2,000
Industry 2	$\frac{1}{3} \times 2,000$	$\frac{1}{25} \times 2,500$	2,500

Therefore final demand which can be met by Industry 1 is equal to $2,000 - (\frac{2}{15} \times 2,000 + \frac{3}{25} \times 2,500) = 1,433\frac{1}{3}$ and by Industry 2 to $2,500 - (\frac{1}{3} \times 2,000 + \frac{1}{25} \times 2,500) = 1,733\frac{1}{3}$.

3. (a)

$$\begin{bmatrix} 1-0.2 & -0.4 \\ -0.3 & 1-0.2 \end{bmatrix} \begin{bmatrix} X_1 \\ X_2 \end{bmatrix} = \begin{bmatrix} 1,000 \\ 2,000 \end{bmatrix}$$

which can be solved using the inverse matrix to give $X_1 = 3,077$ and $X_2 = 3,654$ approximately.

(b)
$$\begin{bmatrix} 1 & -0.6 \\ -0.4 & 1 \end{bmatrix} \begin{bmatrix} X_1 \\ X_2 \end{bmatrix} = \begin{bmatrix} 1,000 \\ 2,000 \end{bmatrix}$$

$X_1 = 2,895$ and $X_2 = 3,158$ approximately.

EXERCISES 3.2

1. Let $\qquad\qquad TC = a + bx + cx^2$

   then $\qquad\qquad\qquad 4 = a \qquad\qquad\qquad$ (1)

   $\qquad\qquad\qquad\qquad 14 = a + 2b + 4c \qquad$ (2)

   and $\qquad\qquad\qquad 58 = a + 6b + 36c \qquad$ (3)

   Subtract eq. (1) from eqs. (2) and (3)

   $\qquad\qquad\qquad\qquad 10 = 2b + 4c \qquad\qquad$ (4)

   $\qquad\qquad\qquad\qquad 54 = 6b + 36c \qquad\qquad$ (5)

   Subtract 3 times eq. (4) from eq. (5)

   $\qquad\qquad\qquad\qquad 24 = 24c$

   $\therefore \qquad\qquad\qquad c = 1$

   Substituting this value of c in eq. (4) gives

   $\qquad\qquad\qquad\qquad b = 3$

   $\therefore \qquad\qquad a = 4 \qquad b = 3 \qquad c = 1$

   and $\qquad\qquad TC = 4 + 3x + x^2$

2. $TC = 10 + 2x + \frac{1}{2}x^2$

3. $TC = 25 + \frac{1}{2}x + \frac{1}{4}x^2$

4. $TC = 20 + \frac{4}{5}x + \frac{1}{50}x^2$

5. $TC = 20 + \frac{1}{2}x^2$

EXERCISES 3.6

1. (a) $TC = 50x$

   $$\therefore \quad AC = \frac{50x}{x} = 50$$

   when $x \to \infty$, $AC \to 50$

   (b) when $x \to \infty$, $AC \to 1$
   (c) when $x \to \infty$, $AC \to \infty$

2. (a) $TC = 25 - 6x + x^2$
   (b) 305
   (d) total revenue equals total cost when

   $$10x + 15 = 25 - 6x + x^2$$

   that is when $x^2 - 16x + 10 = 0$

   The roots of this quadratic are given by

   $$x = \frac{16 \pm \sqrt{(16^2 - 4 \times 10)}}{2}$$

   $$= \tfrac{1}{2}(16 \pm \sqrt{(256 - 40)}) = \tfrac{1}{2}(16 \pm \sqrt{(216)})$$

   $$= 8 \pm 7.35 = 15.35 \quad \text{or} \quad 0.65$$

   these are the two levels of output at which total revenue equals total cost.

4. The equations are satisfied by the values $x = 1$ and $x = 15$.

5.
   $$y = 20 + 3x + x^2$$
   $$y = 5x + b$$

   $\therefore$
   $$20 + 3x + x^2 = 5x + b$$

   or
   $$x^2 - 2x + (20 - b) = 0$$

   $$x = \frac{2 \pm \sqrt{[4 - 4(20 - b)]}}{2}$$

   $$= 1 \pm \sqrt{[1 - (20 - b)]}$$

   $$= 1 \pm \sqrt{(b - 19)}$$

$$\text{when} \quad b=20 \quad x=1\pm1=2 \quad \text{or} \quad 0 \quad \text{(2 real solutions)}$$
$$b=19 \quad x=1 \quad \text{(the solutions are coincident)}$$
$$b=18 \quad x=1\pm\sqrt{(-1)} \quad \text{(2 complex solutions)}$$

EXERCISES 3.8

1.
$$2p^2 - 3p - 40 = 250 - 4p - p^2$$

∴
$$3p^2 + p - 290 = 0$$

$$p = \frac{-1 \pm \sqrt{(1 + 12 \times 290)}}{6}$$

$$= \frac{-1 \pm \sqrt{3,481}}{6} = \frac{-1 \pm 59}{6}$$

$$= \frac{58}{6} \quad \text{or} \quad \frac{-60}{6}$$

In this example we take the positive root, $p = 9\frac{2}{3}$, and at this value $q = 117\frac{8}{9}$.

2. (a) $p=5$ $q=10$ (b) $p=4$ $q=4$
   (c) $p=2$ $q=3$

EXERCISE 3.11

1. (a) 6   (b) 31   (c) 36

EXERCISES 4.3

1. (a) 19, 225   (b) 950, 12,750
   (c) −210, −2,250   (d) 82, 1,290

2. £300   3. £1,090, £6,790

4. Arithmetic progression with $20 = a + (n-1)d = 10 + 15d$. Hence $d = \frac{2}{3}$ and rate of interest is $\frac{2}{3}/10 = 0.067$ or 6.7 per cent.

EXERCISES 4.5

1. (a) 2,430; $S(10) = \dfrac{10(1 - 3^{10})}{-2} = 295,240$

(b) $\frac{1}{3}$; $S(10) = \dfrac{81[1-(\frac{1}{3})^{10}]}{\frac{2}{3}} = 121.5$ approximately

$$S(\infty) = \dfrac{81}{1-(\frac{1}{3})} = 243/2 = 121\tfrac{1}{2}$$

(c) -64; $S(10) = \dfrac{2[1-(-2)^{10}]}{3} = -682$

(d) 32; $S(10) = \dfrac{-1,024[1-(-\frac{1}{2})^{10}]}{1-(-\frac{1}{2})} = -682$ approximately

$$S(\infty) = \dfrac{-1,024}{1-(-\frac{1}{2})} = -682\tfrac{2}{3}$$

(e) 0.00001; $S(10) = \dfrac{1[1-(0.1)^{10}]}{1-0.1} = 1.111$ approximately

$$S(\infty) = \dfrac{1}{1-0.1} = \tfrac{10}{9}$$

2. (a) $300(1.05)^{15} = 624$ (b) $300(1.10)^{15} = 1,255$

3. $S(25) = \dfrac{150(1-1.06^{25})}{(1-1.06)} = 8,225$

4. $4(1+r)^4 = 5$

$\therefore$ $(1+r)^4 = 1.25$

$\therefore$ $1+r = 1.06$ approximately

and implied rate of compound interest is 6 per cent.

EXERCISES 4.7

1. (a) Present value of £800 received in 10 years time is given by £800$/(1+i)^{10}$; this is equal to £300.

$\therefore$ $(1+i)^{10} = 800/300 = \tfrac{8}{3}$

We therefore require the discount factors which make

$$\dfrac{1}{(1+i)^{10}} = \tfrac{3}{8} = 0.375$$

From Table 4.5 it can be seen that 0.386 is the factor used to discount sums of money received 10 years hence when the interest rate is equal to 10 per cent. Therefore the implied rate of compound interest is slightly greater than 10 per cent.

(b) The present value of the £800 when the interest rate is 16 per cent is given by £800 × 0.227 = £186.6

2. (a) £1,000 × 0.270 = £270   (b) £1,000 × 0.463 = £463

3. (a) £250 × 0.621 = £155¼
   (b) £155¼/0.751 = £207 approximately

4. (a) Present value of $A = 100 \times 0.909 + 200 \times 0.826 + 300 \times 0.751$
$$= 481.4$$

   Present value of $B = 150 \times 0.909 + 300 \times 0.826 + 100 \times 0.751$
$$= 459.3$$

   ∴ Project A has the greater present value if the discount rate is 10 per cent.

5. Let r = internal rate of return; then
$$\frac{(100-120)}{(1+r)} + \frac{(110-120)}{(1+r)^2} + \frac{(160-100)}{(1+r)^3} = 0$$

$$\frac{-20(1+r)^2 - 10(1+r) + 60}{(1+r)^3} = 0$$

∴
$$2(1+r)^2 + (1+r) - 6 = 0$$
$$2 + 4r + 2r^2 + 1 + r - 6 = 0$$
$$2r^2 + 5r - 3 = 0$$
$$r = \frac{-5 \pm \sqrt{(25+24)}}{4} = \frac{-5 \pm 7}{4}$$

Taking the positive root, $r = 0.5$.

EXERCISES 4.9

1. $PV = \dfrac{100}{0.08}\left[1 - \dfrac{1}{1.08^{20}}\right] = 981$

2. $PV = 25/0.04 = 625$

3. $A = \dfrac{2{,}500(1 - 1.09)}{1 - (1.09)^{15}} = 85.25$

4. (a) $PV = \dfrac{6}{0.06}\left[1 - \dfrac{1}{(1.06)^{10}}\right] = 44.16$

   (b) $PV = \dfrac{6}{0.06} = 100$

EXERCISES 4.13

1. (a) Simple interest of 6 per cent per annum produces £45
   Compound interest of 5 per cent produces £41
   Compound interest of 4 per cent paid twice per
   year produces £33
   ∴ most profitable interest is that returning simple interest
   at 6 per cent.

   (b) £90, £94, £73; most profitable investment is that returning
   compound interest of 5 per cent per annum.

2. $50(1.02)^{12} = 63.4$

3. (a) $1 + 5x + 10x^2 + 10x^3 + 5x^4 + x^5$

   (b) $8 - 12x + 6x^2 - x^3$

   (c) $81 + 108\left(\dfrac{1}{x}\right) + 54\left(\dfrac{1}{x}\right)^2 + 12\left(\dfrac{1}{x}\right)^3 + \left(\dfrac{1}{x}\right)^4$

   (d) $64 - 192x^2 + 240x^4 - 160x^6 + 60x^8 - 12x^{10} + x^{12}$

4. (a) $e^{0.1} = 1 + 0.1 + \dfrac{(0.1)^2}{2} + \dfrac{(0.1)^3}{6} + \cdots$

   $= 1 + 0.1 + 0.005 + 0.00017 + \cdots$

   $= 1.10517$ approximately (from tables $e^{0.1} = 1.1052$)

   (b) $e^{0.5} = 1 + 0.5 + \dfrac{(0.5)^2}{2} + \dfrac{(0.5)^3}{6} + \cdots$

   $= 1.6458$ approximately (from tables $e^{0.5} = 1.6487$)

MATHEMATICS FOR ECONOMISTS

If we include the next term in the expansion we obtain a closer approximation with $e^{0.5} = 1.6484$.

(c) $e^2 = 1 + 2 + \frac{4}{2} + \frac{8}{6} + \frac{16}{24} + \frac{32}{120} + \frac{64}{720} + \cdots$

   $= 7.35556$ approximately (from tables $e^2 = 7.3891$)

$\therefore$ in this case we need to include more terms to obtain a good approximation.

(d) $e^{-1} = 1 - 1 + \frac{1}{2} - \frac{1}{6} + \frac{1}{24} - \frac{1}{120} + \frac{1}{720} - \cdots$

   $= 0.368$ approximately (from tables $e^{-1} = 0.3679$).

EXERCISES 5.3

1. $y = 3x - 4$, $dy/dx = 3$

2. $y = x^2 - 3x + 3$, $dy/dx = 2x - 3$

3. $12x^3 - 3x^2 + 2x + 25$ 4. $18x^2 - x^3 + x^2 + 2/x^3$

5. $(1/x - 3x + 2)(4x + 3) + (2x^2 + 3x - 1)(-1/x^2 - 3)$

6. $(3x + 2/x^2)(2) + (2x + 4)(3 - 4/x^3)$

7. $\dfrac{(x^2 + 3x + 2)(8x) - (4x^2 + 4)(2x + 3)}{(x^2 + 3x + 2)^2}$

8. $5(2x + 3)^4 \cdot 2$ 9. $\dfrac{1}{2x + 3} \cdot 2$

10. $\log (2x^2 + 3x - 5) + \dfrac{x(4x + 3)}{2x^2 + 3x - 5}$

11. $6e^{2x} - 12e^{-3x} + 2xe^x + x^2 e^x$

12. $2 \cos 2x - 15 \sin 5x - \dfrac{3}{\cos^2 3x}$

EXERCISES 5.6

1. (a) $TC = 4q^3 + 2q^2 - 25q$

   Marginal cost $(MC) = \dfrac{d(TC)}{dq} = 12q^2 + 4q - 25$

310

Average cost $(AC) = \dfrac{TC}{q} = 4q^2 + 2q - 25$

(b) $MC = (q^3 - 3q)5 + (16 + 5q)(3q^2 - 3)$

$AC = (q^2 - 3)(16 + 5q)$

(c) $MC = e^{2q}(6) + 12qe^{2q}$

$AC = \dfrac{25}{q} + 6e^{2q}$

(d) $MC = (q^2 - q)\left(\dfrac{3}{q}\right) + (3 \log q + 5)(2q - 1)$

$AC = (3 \log q + 5)(q - 1)$

2. Elasticity of demand $= \dfrac{dq}{dp}\dfrac{p}{q}$

(a) $p + 2q = 50$ $\therefore$ $dq/dp = -\frac{1}{2}$

and when $p = 10$, $q = 20$

$\therefore$ elasticity of demand $= -\left(\frac{1}{2}\right)\left(\frac{10}{20}\right) = -\frac{1}{4}$

(b) -2 (c) -6

3. $q = cp^{-1}$ $\therefore$ $\dfrac{dq}{dp} = -cp^{-2} = -\dfrac{c}{p^2}$

and elasticity of demand $= \left(-\dfrac{c}{p^2}\right)\left(\dfrac{p}{q}\right) = -\dfrac{c}{pq}$

but $pq = c$ $\therefore$ elasticity of demand $= -1$.

Similarly if $q = cp^{-4}$

elasticity of demand $= (-4cp^{-5})\left(\dfrac{p}{q}\right) = -\dfrac{4cp^{-4}}{q} = -4$

and in general if $q = cp^{-n}$

elasticity of demand $= (-ncp^{-(n+1)})\left(\dfrac{p}{q}\right) = -\dfrac{ncp^{-n}}{q} = -n$

EXERCISES 5.9

1. (a) $y = 3x^2 - 120x + 30$

$$\frac{dy}{dx} = 6x - 120$$

For a stationary value $\frac{dy}{dx} = 0$

so $6x - 120 = 0$ and $x = 20$

$$\frac{d^2y}{dx^2} = 6 \text{ which is positive}$$

$\therefore$ $x = 20$ is a minimum value and $y = -1170$

(b) $y = 16 - 8x - x^2$

$$\frac{dy}{dx} = -8 - 2x \qquad \frac{d^2y}{dx^2} = -2$$

$dy/dx = 0$ when $x = -4$ and there is a maximum value at this point with $y = 32$

(c) $y = x^4$

$$\frac{dy}{dx} = 4x^3 \qquad \frac{d^2y}{dx^2} = 12x^2$$

There is a stationary value when $4x^3 = 0$, that is $x = 0$. This is in fact a minimum, with $y = 0$, and the curve is symmetrical about the y-axis.

(d) $y = (x - 5)^3$

$$\frac{dy}{dx} = 3(x - 5)^2 \qquad \frac{d^2y}{dx^2} = 6(x - 5)$$

There is a stationary value when

$3(x - 5)^2 = 0$ that is when $x = 5$

at this point d^2y/dx^2 is also equal to 0 and it is necessary to sketch the curve in order to see whether it is a maximum, minimum or point of inflexion.

2. $C = 500 + 4q + \frac{1}{2}q^2$

$\therefore$ average cost $= \dfrac{500}{q} + 4 + \dfrac{q}{2}$

$\dfrac{d(AC)}{dq} = \dfrac{-500}{q^2} + \frac{1}{2}$ and $\dfrac{d^2(AC)}{dq^2} = \dfrac{1000}{q^3}$

There is a stationary point when $500/q^2 = \frac{1}{2}$

that is $q^2 = 1,000$ and $q = \pm 31.6$

There is a minimum when $q = 31.6$ because $d^2AC/dq^2 > 0$ for this value.

3. Demand function is given by $q + 2p = 10$

Total revenue $= pq = p(10 - 2p)$

$\therefore \dfrac{d(TR)}{dp} = 10 - 4p,$ $\dfrac{d^2(TR)}{dp^2} = -4$

This has a stationary value when $p = \frac{10}{4} = \frac{5}{2}$

The second-order derivative is negative and there is therefore a maximum when $p = \frac{5}{2}$.

$$q = 10 - 2p = 10 - 5 = 5$$

Elasticity of demand $= \dfrac{dq}{dp}\dfrac{p}{q} = \dfrac{(-2)(\frac{5}{2})}{5} = -1$

4. Net revenue $= pq - TC = q(12 - \frac{1}{2}q) - (q^3 - 20q^2 + 20)$

$\therefore \dfrac{d(NR)}{dq} = 12 - q - 3q^2 + 40q = -3q^2 + 39q + 12$

This has a stationary value when $q^2 - 13q - 4 = 0$, that is when $q = 13.3$ or -0.3

$$\dfrac{d^2(NR)}{dq^2} = -6q + 39$$

This is positive when $q = -0.3$ and negative when $q = 13.3$.

$\therefore$ net revenue is maximised when $q = 13.3$ and $p = 5.35$.

313

EXERCISES 5.14

1. Taylor's theorem states

$$f(a+x)=f(a)+xf'(a)+\frac{x^2}{2!}f''(a)+\cdots$$

Let $\qquad f(a+x)=(1+x)^4$

$$f(a)=1$$

$$f'(a+x)= 4(1+x)^3 \qquad \text{and} \qquad f'(a)=4$$

$$f''(a+x)=12(1+x)^2 \qquad \text{and} \qquad f''(a)=12$$

$$f'''(a+x)=24(1+x) \qquad \text{and} \qquad f'''(a)=24$$

$$f^4(a+x)=24 \qquad\qquad \text{and} \qquad f^4(a)=24$$

$$\therefore \quad (1+x)^4=1+x(4)+\frac{x^2}{2}(12)+\frac{x^3}{6}(24)+\frac{x^4}{24}(24)$$

$$=1+4x+6x^2+4x^3+x^4$$

and $1.5^4=(1+\frac{1}{2})^4=1+4\times\frac{1}{2}+6\times\frac{1}{4}+4\times\frac{1}{8}+\frac{1}{16}$

$$=1+2+1.5+0.5+0.0625=5.0625$$

2. MacLaurin's theorem states

$$f(x)=f(0)+xf'(0)+\frac{x^2}{2!}f''(0)+\cdots$$

Let $\qquad f(x)=(1+x)^{-1}$

then $\qquad f'(x)=-(1+x)^{-2}$

$$f''(x)=2(1+x)^{-3}$$

$$f'''(x)=-6(1+x)^{-4}$$

$$\vdots$$

Therefore $(1+x)^{-1}=1-x+x^2-x^3+\cdots$

3. (a) $f(x)=x^2+4.55x-8.70$

$\qquad f'(x)=2x+4.55$

Using as a starting point $a = 1.5$

$f(1.5) = 2.25 + 6.825 - 8.70 = 0.375$

$f'(1.5) = 3.0 + 4.55 = 7.55$

$$\therefore \quad h_1 = -\frac{f(a)}{f'(a)} = -\frac{0.375}{7.55} = -0.0497$$

$\therefore$ new approximate root is $1.5 - 0.05 = 1.45$ for which

$$f(1.45) = 2.1 + 6.6 - 8.7 = 0$$

This result shows that 1.45 is a good approximation to one of the values of x which satisfies the equation.

(b) $f(x) = x^3 - 2.34x^2 + 2x - 4.68$

$f'(x) = 3x^2 - 4.68x + 2$

Using as a starting point $a = 2.25$

$f(a) = 11.39 - 11.85 + 4.50 - 4.68 = -0.64$

$f'(a) = 15.19 - 10.53 + 2 = 6.66$

$$\therefore \quad h_1 = -\frac{f(a)}{f'(a)} = \frac{0.64}{6.66} = 0.096$$

$\therefore$ new approximate root is $2.25 + 0.096 = 2.346$

for which $f(2.346) = 0.045$. A closer approximation can be obtained by repeating this process.

(c) $x = 4.1$ for which $f(4.1) = 0$.

4. $PV_1 = \dfrac{70}{R} + \dfrac{80}{R^2} + \dfrac{100}{R^3} + \dfrac{100}{R^4}$

$PV_2 = \dfrac{60}{R} + \dfrac{90}{R^2} + \dfrac{80}{R^3} + \dfrac{130}{R^4}$

these are equal when $PV_1 - PV_2 = 0$
that is

$$\frac{10}{R} - \frac{10}{R^2} + \frac{20}{R^3} - \frac{30}{R^4} = 0$$

or
$$R^3 - R^2 + 2R - 3 = 0$$
$$f(R) = R^3 - R^2 + 2R - 3$$
$$f'(R) = 3R^2 - 2R + 2$$

Using Newton's method with $a = 1$ as a starting point, we obtain
$$f(a) = 1 - 1 + 2 - 3 = -1$$
$$f'(a) = 3 - 2 + 2 = 3$$

$\therefore$
$$h_1 = -\frac{f(a)}{f'(a)} = \tfrac{1}{3} = 0.33$$

$\therefore$ approximate root is $1 + 0.33 = 1.33$; using this as a new starting point, we have

$$f(1.33) = 0.2437$$
$$f'(1.33) = 4.6467$$

$\therefore$
$$h_2 = -\frac{f(1.33)}{f'(1.33)} = -\frac{0.2437}{4.6467} = -0.052$$

Try $R = 1.33 - 0.052 = 1.278$ as the next root: $f(1.278) = 0.101$ and so the rate of interest is approximately 27.8 per cent. A more accurate value can be obtained by repeating the process.

EXERCISE 6.2

1. (a) $\dfrac{dTC}{dq} = 1$ $\quad \therefore \quad TC = q + C$

   when $q = 0$, $\qquad TC = 600$

   $\therefore \quad TC = 600 + q$

   (b) $TC = 50 + q$

   (c) $TC = x^2 + 3x + C$

   $100 = 25 + 15 + C$

   $\therefore \quad C = 60$ $\qquad$ and $\qquad TC = x^2 + 3x + 60.$

EXERCISES 6.4

1. (a) $TC = x^2 + 3x + 10$
   (b) $TC = \frac{1}{3}x^3 + x^2 + 4x + 90$

2. (a) $\frac{3}{2}x^2 - 4x + C$ (b) $\frac{2}{3}x^3 - 2x^2 + 3x + C$
   (c) $\frac{1}{2}x^6 + 4/x + x^2 + C$ (d) $-1/x^2 - 4 \log x + C$
   (e) $\log (3x - 5) + C$ (f) $\log (2x^2 + x - 4) + C$
   (g) $\frac{1}{4} \log (x^4 - 2x^2 + 2) + C$ (h) $2 \log (x^2 + 3x + 1)$
   (i) $\frac{2}{3}e^{3x} - e^{-x} + C$ (j) $2e^x - \frac{4}{3}e^{3x} + C$
   (k) $-2 \cos x - 4 \sin x$

EXERCISE 6.6

1. (a) $\int\limits_{x_1=0}^{x_2=3} y \, dx = \int\limits_{0}^{3} 2x \, dx = [x^2]_0^3 = 9$

   (b) $[x^2]_3^6 = 36 - 9 = 27$

   (c) $[x^2]_0^6 = 36$

   (d) $[\frac{5}{2}x^2]_6^8 = \frac{5}{2}[64 - 36] = 70$

   (e) $[2x + \frac{3}{2}x^2]_0^3 = 6 + \frac{27}{2} = 19\frac{1}{2}$

EXERCISE 6.8

1. (a) $y = 2x$ crosses the x-axis at $x = 0$.

   $\therefore$ area $= 2 \int\limits_{0}^{1} 2x \, dx = 2[x^2]_0^1 = 2$

   (b) $y = 2 + 3x$ crosses the x-axis at $x = -\frac{2}{3}$

   $\therefore$ area $= - \int\limits_{-3}^{-2/3} (2 + 3x) \, dx + \int\limits_{-2/3}^{2} (2 + 3x) \, dx$

   $= - [2x + \frac{3}{2}x^2]_{-3}^{-2/3} + [2x + \frac{3}{2}x^2]_{-2/3}^{2}$

   $= - [-\frac{4}{3} + \frac{2}{3} - (-6 + 13.5)] + [4 + 6 - (-\frac{4}{3} + \frac{2}{3})]$

   $= 8.16 + 10.67 = 18.83$

317

(c) $\text{Area} = \int_{-2}^{3} x^2 \, dx = [\frac{1}{3}x^3]_{-2}^{3} = \frac{8}{3} + 9 = 11\frac{2}{3}$

(d) $y = 1 + x + x^2$ does not cross the x-axis

$\therefore \quad \text{area} = \int_{-4}^{+1} (1 + x + x^2) \, dx$

$\qquad = [x + \frac{1}{2}x^2 + \frac{1}{3}x^3]_{-4}^{+1} = 115/6$

EXERCISES 6.11

1. Let $(5x^2 + 4) = u$ and $e^{3x} \, dx = dv$

Then $\int (5x^2 + 4)e^{3x} \, dx = (5x^2 + 4)\frac{1}{3}e^{3x} - \int \dfrac{10xe^{3x}}{3} \, dx$

and $\quad \int 10xe^{3x} \, dx = 10x(\frac{1}{3}e^{3x}) - \int \dfrac{10e^{3x}}{3} \, dx$

and $\quad \int 10e^{3x} \, dx = 10(\frac{1}{3}e^{3x})$

$\therefore \quad \int (5x^2 + 4)e^{3x} \, dx = e^{3x}\left[\dfrac{5x^2 + 4}{3} - \dfrac{10x}{9} + \dfrac{10}{27}\right] + C$

2. Let $\log x = u$ and $x^2 \, dx = dv$

Then $\int x^2 \log x \, dx = \int \log x \, x^2 \, dx$

$\qquad\qquad\qquad = \log x(\frac{1}{3}x^3) - \int \frac{1}{3}x^3 \dfrac{1}{x} \, dx$

$\qquad\qquad\qquad = \frac{1}{3}x^3 \log x - \frac{1}{3} \int x^2 \, dx$

$\qquad\qquad\qquad = \frac{1}{3}[x^3 \log x - \frac{1}{3}x^3] + C$

$\qquad\qquad\qquad = \frac{1}{3}x^3[\log x - \frac{1}{3}] + C$

3. Let $\dfrac{3}{(x+1)(x-1)} = \dfrac{A}{(x+1)} + \dfrac{B}{(x-1)}$

Then $\qquad\qquad 3 = A(x-1) + B(x+1)$

and $\qquad\qquad 3 = -A + B$

$\qquad\qquad\qquad 0 = A + B$

$$\therefore \quad A = -\tfrac{3}{2} \qquad B = \tfrac{3}{2}$$

$$\therefore \quad \int \frac{3}{(x+1)(x-1)} \, dx = \int \frac{-\tfrac{3}{2}}{(x+1)} \, dx + \int \frac{\tfrac{3}{2}}{(x-1)} \, dx$$

$$= -\tfrac{3}{2} \log (x+1) + \tfrac{3}{2} \log (x-1) + C$$

$$= \tfrac{3}{2} [\log (x-1) - \log (x+1)] + C$$

$$= \tfrac{3}{2} \log \frac{(x-1)}{(x+1)} + C$$

4. Let
$$\frac{x}{(x-2)(x^2-3)} = \frac{A}{(x-2)} + \frac{Bx+C}{(x^2-3)}$$

then
$$x = A(x^2-3) + (Bx+C)(x-2)$$
$$= Ax^2 - 3A + Bx^2 + Cx - 2Bx - 2C$$

$$\therefore \quad A + B = 0$$
$$-2B + C = 1$$
$$-3A - 2C = 0$$

and $\quad A = 2, \qquad B = -2, \qquad C = -3.$

$$\therefore \quad \int \frac{x}{(x-2)(x^2-3)} \, dx = \int \frac{2}{(x-2)} \, dx - \int \frac{2x+3}{(x^2-3)} \, dx$$

$$= 2 \log (x-2) - \log (x^2-3) - 3 \int \frac{1}{x^2-3} \, dx$$

$$= \log \left[\frac{(x-2)^2}{(x^2-3)} \right] + \frac{3}{2\sqrt{3}} \log \left(\frac{\sqrt{3}+x}{\sqrt{3}-x} \right) + D$$

This can be further simplified if necessary.

5. Let $4x + 3 = u$; then $du = 4 \, dx$

$$\therefore \quad \int (4x+3)^{10} \, dx = \int \frac{u^{10}}{4} \, du = \frac{u^{11}}{44} + C = \frac{(4x+3)^{11}}{44} + C$$

6. Let $\log x = u$ and $dx = dv$

Then $\int \log x \, dx = (\log x)x - \int x \frac{1}{x} \, dx$

$$= x \log x - x + C$$

$$= x(\log x - 1) + C$$

EXERCISES 7.4

1. (a) $\dfrac{\partial z}{\partial x} = y + 2xy + y^2$

$\dfrac{\partial z}{\partial y} = x + x^2 + 2xy$

(b) $\dfrac{\partial z}{\partial x} = 6xy + 4y^2 + 6y$

$\dfrac{\partial z}{\partial y} = 3x^2 + 8xy + 6x$

(c) $\dfrac{\partial z}{\partial x} = 2x \sin y - y^2 \sin x$

$\dfrac{\partial z}{\partial y} = x^2 \cos y + 2y \cos x$

(d) $\dfrac{\partial z}{\partial x} = (x + y)e^{x+y} + e^{x+y}$

$\dfrac{\partial z}{\partial y} = (x + y)e^{x+y} + e^{x+y}$

(e) $\dfrac{\partial z}{\partial x} = \dfrac{2x}{x^2 + y^2}$

$\dfrac{\partial z}{\partial y} = \dfrac{2y}{x^2 + y^2}$

(f) $\dfrac{\partial z}{\partial x} = \dfrac{(x - y) - (x + y)}{(x - y)^2} = \dfrac{-2y}{(x - y)^2}$

$\dfrac{\partial z}{\partial y} = \dfrac{(x - y) - (x + y)(-1)}{(x - y)^2} = \dfrac{2x}{(x - y)^2}$

(g) $\dfrac{\partial z}{\partial x} = \dfrac{(y + \sin x) - (x + \sin y) \cos x}{(y + \sin x)^2}$

$\dfrac{\partial z}{\partial y} = \dfrac{(y + \sin x) \cos y - (x + \sin y)}{(y + \sin x)^2}$

2. Marginal product of capital is

$$\frac{\partial q}{\partial C} = 4(C-30) + 2(L-20)$$

$$= 80 + 20 = 100 \text{ when } C = 50, \ L = 30$$

Marginal product of labour is

$$\frac{\partial q}{\partial L} = 6(L-20) + 2(C-30) = 60 + 40 = 100$$

when $C = 50$ and $L = 30$.

3. (a) Partial elasticity of demand for apples with respect to price of apples is

$$\frac{\partial q}{\partial p_a} \frac{p_a}{q} = \frac{(-2p_a - p_o)p_a}{q} = -\frac{14(5)}{219} = -\frac{70}{219}$$

when $p_a = 5$, $p_o = 4$ and $q = 219$.

(b) Partial elasticity of demand for apples with respect to price of oranges is

$$\frac{\partial q}{\partial p_o} \frac{p_o}{q} = \frac{(6-p_a)p_o}{q} = \frac{4}{219}$$

when $p_a = 5$, $p_o = 4$ and $q = 219$.

EXERCISES 7.7

1. $dq = \dfrac{\partial q}{\partial L} dL + \dfrac{\partial q}{\partial C} dC$

$$= (3L^2 - 3 + 4C) \, dL + (4L - 2C) \, dC$$

$$= 317(1) + 30(1) = 347 \text{ when } L = 10 \text{ and } C = 5$$

If $dC = 0$ and $dL = 1$, $dq = 317$.

2. $dq = \dfrac{\partial q}{\partial p_a} dp_a + \dfrac{\partial q}{\partial p_o} dp_o$

$$= (-2p_a - p_o) \, dp_a + (6 - p_a) \, dp_o$$

$$= -14(1) + 1(1) = -13$$

3. $\dfrac{dz}{dt} = \dfrac{\partial z}{\partial x}\dfrac{dx}{dt} + \dfrac{\partial z}{\partial y}\dfrac{dy}{dt}$

(a) $\dfrac{dx}{dt} = e^t(-\sin t) + e^t \cos t, \qquad \dfrac{dy}{dt} = e^t \cos t + e^t \sin t$

$\dfrac{dz}{dt} = 2x(e^t)(\cos t - \sin t) + (-2y)(e^t)(\cos t + \sin t)$

(b) $\dfrac{dx}{dt} = 2te^t + t^2 e^t, \qquad \dfrac{dy}{dt} = \cos t$

$\dfrac{dz}{dt} = \left(\dfrac{1}{x+y}\right)(2t + t^2)e^t + \left(\dfrac{1}{x+y}\right)\cos t$

(c) $\dfrac{dx}{dt} = e^t, \qquad \dfrac{dy}{dt} = \dfrac{1}{t}$

$\dfrac{dz}{dt} = (2xye^{x^2})e^t + (e^{x^2})\left(\dfrac{1}{t}\right)$

(d) $\dfrac{dx}{dt} = 1 - \dfrac{1}{t^2}, \qquad \dfrac{dy}{dt} = 1 + \dfrac{1}{t^2}$

$\dfrac{dz}{dt} = (2x+y)\left(1 - \dfrac{1}{t^2}\right) + (x+2y)\left(1 + \dfrac{1}{t^2}\right)$

EXERCISES 7.9

1. $\dfrac{dy}{dx} = -\dfrac{(\partial z/\partial x)}{(\partial z/\partial y)}$

(a) $\dfrac{dy}{dx} = -\dfrac{(-8xy^2 + 12y + 4)}{3y^2 - 8x^2 y + 6x^2}$

(b) $\dfrac{dy}{dx} = -\dfrac{(2xy^2 + e^x)}{2x^2 y - e^y}$

(c) $\dfrac{dy}{dx} = -\dfrac{[(3/(x+y)] + 8x}{[(3/(x+y)] - 2y}$

(d) $\dfrac{dy}{dx} = -\dfrac{[\cos(x+y) - 3y - 10y]}{\cos(x+y) - 3x - 10x}$

2. When $p = 2$, $q = 12.18$.

By differentiating term by term

$$2q\frac{dq}{dp} = \frac{(2+p)(-2p) - (500-p^2)}{(2+p)^2} + p\frac{dq}{dp} + q$$

Hence $\dfrac{dq}{dp} = \dfrac{1}{(2q-p)}\left[\dfrac{-4p - 500 - p^2}{(2+p)^2} + q\right]$

and elasticity when $p = 2$ and $q = 12.18$ is

$$\frac{p}{q}\frac{dq}{dp} = \frac{2}{12.18}\left(\frac{1}{22.36}\right)\left[-\frac{512}{16} + 12.18\right] = -0.15$$

EXERCISES 7.12

1. Replace x and y by tx and ty and show $f(tx, ty) = t^k f(x, y)$ where k is the degree of homogeneity: (a) $k = 3$; (b) $k = -1$; (c) $k = 3$.

2. Euler's theorem: $xz_x + yz_y = kz$

   For (a), $z_x = 3x^2 - 2xy$, $z_y = 9y^2 - x^2$

   hence $x(3x^2 - 2xy) + y(9y^2 - x^2) = 3f(x, y)$

   For (b), $z_x = \dfrac{(4x^3)y - 12x^2(xy + y^2)}{(4x^3)^2}$, $\qquad z_y = \dfrac{x + 2y}{4x^3}$

   hence $xz_x + yz_y = -f(x, y)$

3. Own price elasticity is

$$\frac{p}{q}\frac{\partial q}{\partial p} = \frac{p}{q}(4r - 4p)$$

Cross elasticity is

$$\frac{r}{q}\frac{\partial q}{\partial r} = \frac{r}{q}(20r + 4p)$$

Sum is 2, the degree of homogeneity.

4. Replace L, C by tL, tC, then the first one gives $t^{0.8}q$ and the second gives $t^{1.1}q$ as required.

EXERCISES 7.14

1. (a) $z = x^2 - xy + y^2$

$z_x = 2x - y$ $\qquad z_y = -x + 2y$

$z_{xx} = 2$ $\qquad\qquad z_{yy} = 2$

$z_{xy} = -1$

(b) $z = (x + y)^3$

$z_x = 3(x + y)^2$ $\qquad z_y = 3(x + y)^2$

$z_{xx} = 6(x + y)$ $\qquad z_{yy} = 6(x + y)$

$z_{xy} = 6(x + y)$

2. Find when $z_x = 0 = z_y$ and check for

Maximum: $\qquad z_{xx} z_{yy} > (z_{xy})^2$, $\qquad z_{xx} < 0$, $\qquad z_{yy} < 0$

Minimum: $\qquad z_{xx} z_{yy} > (z_{xy})^2$, $\qquad z_{xx} > 0$, $\qquad z_{yy} > 0$

Saddle point: $z_{xx} z_{yy} < (z_{xy})^2$, $\qquad z_{xx} = 0$

(a) $z_x = 0 = z_y$ when $x = 0$, $y = 0$ and $x = \frac{2}{3}$, $y = -\frac{2}{3}$

$z_{xx} = 6x$, $\qquad z_{yy} = 2$, $\qquad z_{xy} = 2$

For $x = 0$, $y = 0$, tests fail—further investigation needed.

For $x = \frac{2}{3}$, $y = -\frac{2}{3}$, minimum with $z = -\dfrac{4}{27}$.

(b) $z_x = 0 = z_y$ when $x = \frac{1}{2}$ and $y = \frac{1}{4}$. Maximum value is $z = 1$.

3. Method is as in Problem 2 above.

(a) $z_x = 0 = z_y$ when $x = 2$ and $y = 0.5$.

$z_{xy} = 0$, $z_{xx} = 2$, $z_{yy} = 2$ so minimum value of $z = 45.75$.

(b) $z_x = 0 = z_y$ when $x = \pm \sqrt{(0.67)}$, $y = 1.5$.

Minimum is $z = 196.7$ when $x = \sqrt{(0.67)}$, $y = 1.5$

(c) $z_x = 0 = z_y$ when $x = 3$, $y = 4$ and $x = -3$, $y = 4$.

Minimum is $z = 180$ when $x = 3$, $y = 4$.

EXERCISES 7.17

1. Using a Lagrangian multiplier, as in Section 7.16,

$z' = 100x_1 - 2x_1^2 + 60x_2 - x_2^2 + \lambda(x_1 + x_2 - 41)$

Setting $z_1' = 0$, $z_2' = 0$, $z_\lambda' = 0$ gives $x_1 = 20.33$

$x_2 = 20.67$ which are the maximising values, $z = 2019.3$ (compare $z = 2000$ when the constraint was $x_1 + x_2 = 40$).

Alternatively, substitute $x_1 = 41 - x_2$ in the expression for z and maximise as in Section 5.7.

2. For the unrestricted case, $z_x = 0 = z_y$ gives $x = 2$, $y = 7$ and $z = 89$ as the minimum value (see Section 7.13 for method).
   For the restricted case, using a Lagrangian multiplier, the first-order partial derivatives are zero when $x = 10/7$, $y = 43/7$, and the minimum value of z is $635/7$

3. (a) Using a Lagrangian multiplier shows that the first-order partial derivatives are zero when $x = 200/41$, $y = 160/41$, and the minimum value of z is $1600/41$.
   (b) Using a Lagrangian multiplier shows that the first-order partial derivatives are zero when $x = 48/17$, $y = 36/17$, and the maximum value of z is $162/17$.
   (c) Using a Lagrangian multiplier shows that the first-order partial derivatives set equal to zero give

   $$0 = 6xy^2 + \lambda, \qquad 0 = 6x^2y + 3\lambda, \qquad 0 = x + 3y - 18$$

   These are satisfied by three pairs of values:

$x = 0$,	$y = 6$,	minimum value $z = 0$
$x = 18$,	$y = 0$,	minimum value $z = 0$
$x = 9$,	$y = 3$,	maximum value $z = 2187$

4. The constraint is $4l + 5c = 28$. Using a Lagrangian multiplier, we find that the first-order partial derivatives are zero when $l = 93/49$ and $c = 200/49$. The maximum value of x is 144.0.

EXERCISES 8.8

1. The situation shown in Table 1 can be expressed as

$$2x_1 + 3x_2 \leqslant 25$$

$$4x_1 + x_2 \leqslant 35$$

where x_1 and x_2 are the numbers of units of product A and B which are manufactured. x_1 and x_2 must both therefore be $\geqslant 0$

The contribution is given by $9x_1 + 7x_2$.

TABLE 1

	Number of hours required per unit of product		Total number of hours available
	A	B	
Machine X	2	3	25
Machine Y	4	1	35
Contribution per unit of product	9	7	

Slack variables x_3 and x_4 are inserted in the equations in order to obtain a first feasible solution. The Simplex tableau is as shown in Tableau 1.

Tableau 1

	x_1	x_2	x_3	x_4	p	p_i/a_{ij}	
x_3	2	3	1	0	25	25/2	
x_4	4	1	0	1	35	35/4	←departing variable
z	−9	−7	0	0	0		
	↑						

entering
variable

Using the method of row and column operations successive tableaux are obtained as shown in Tableaux 2 and 3.

The final tableau shows that the maximum contribution is obtained by manufacturing 8 units of product A and 3 units of product B, in which case $z = 93$.

Check $8 \times 9 + 3 \times 7 = 72 + 21 = 93$.

Tableau 2

	x_1	x_2	x_3	x_4	p	p/a_{ij}	
x_3	0	$\frac{5}{2}$	1	$-\frac{1}{2}$	$\frac{15}{2}$	$3\leftarrow$	departing variable
x_1	1	$\frac{1}{4}$	0	$\frac{1}{4}$	$\frac{35}{4}$	35	
z	0	$-\frac{19}{4}$	0	$\frac{9}{4}$	$\frac{315}{4}$		

entering variable

Tableau 3

	x_1	x_2	x_3	x_4	p
x_2	0	1	$\frac{2}{5}$	$-\frac{1}{5}$	3
x_1	1	0	$-\frac{1}{10}$	$\frac{3}{10}$	8
z	0	0	$\frac{19}{10}$	$\frac{26}{20}$	93

The value of increasing capacity on machine X by one unit is $19/10$ and that on machine Y is $26/20$. These values are obtained directly from the bottom line of the final tableau. They can be verified by using the simplex method with the constraints adjusted by one unit of capacity for machines X and Y separately.

2. (a) (i) $A = 9\frac{1}{2}$, $B = 2$.   (ii) $A = 11$, $B = 1$

   (b) $A = 0$, $B = 8\frac{1}{3}$ (c) $A = 7\frac{1}{2}$, $B = 5$

EXERCISES 9.3

1. Let $Y = VX$, so that $dY = V\,dX + X\,dV$.

Substituting and re-arranging gives

$$(a_1V - b_1 - a_2V^2 + b_2V) \, dX = (a_2VX - b_2X) \, dV$$

or
$$\frac{dX}{X} = \frac{(a_2V - b_2) \, dV}{a_1V - b_1 - a_2V^2 + b_2V}$$

which can be integrated given the values of a_1, a_2, b_1 and b_2.

2. Let q be the quantity demanded at price p. Then we require

$$\frac{p}{q} \frac{dq}{dp} = -n$$

or
$$\frac{dq}{q} = -n \frac{dp}{p}$$

Integrating, we obtain

$$\log q = -n \log p + A' \qquad \text{or} \qquad q = Ap^{-n}$$

Thus $qp^n = A$ is the general form required.

3. Put $y = vx$, $dy = v \, dx + x \, dv$; then, on re-arranging,

$$\frac{v \, dv}{24 - 2v^2} = \frac{dx}{x}$$

or
$$-\tfrac{1}{4} \log (24 - 2v^2) = \log x + A'$$

This becomes

$$A = (24 - 2v^2)x^4 = 24x^4 - 2x^2y^2$$

and using $x = 2$, $y = 4$ gives $A = 256$.

4. Variables separable: $c = ax - \dfrac{bx^2}{2} + A$

5. (a) $\dfrac{dp}{dt} = kap \qquad \text{or} \qquad \dfrac{dp}{p} = ka \, dt$

Integrating gives

$$\log p = kat + A' \qquad \text{or} \qquad p = Ae^{kat}$$

(b) $\dfrac{dp}{dt} = k(ap + bp^2) \qquad \text{or} \qquad \dfrac{dp}{p(a + bp)} = k \, dt$

Using partial fractions (see Section 6.10), we have

$$\frac{dp}{ap} - \frac{b\,dp}{a(a+bp)} = k\,dt$$

or

$$\frac{1}{a}\log p - \frac{1}{a}\log(a+bp) = kt + A'$$

$$\frac{p}{(a+bp)} = Ae^{akt}$$

6. Substituting gives

$$b\frac{dY}{dt} = aY \qquad \text{or} \qquad b\frac{dY}{Y} = a\,dt$$

Hence, $\qquad b\log Y = at + A' \qquad$ or $\qquad Y^b = Ae^{at}$

7. (a) $x\,dy = (x-y)\,dx$

Put $y = vx$, $dy = v\,dx + x\,dv$, then $(-1+2v)\,dx = -x\,dv$

or

$$\frac{dx}{x} + \frac{dv}{2v-1} = 0$$

so $\qquad \log x + \frac{1}{2}\log(2v-1) = A' \qquad$ or $\qquad 2xy - x^2 = A$

(b) Let $X = y+1$ and $Y = 2y+x$; then

$$\frac{dY}{dX} = \frac{dY/dx}{dX/dx} = \frac{2(dy/dx)+1}{dy/dx} = \frac{2(X/Y)+1}{X/Y}$$

and $\qquad X\,dY = (2X+Y)\,dX$

Let $Y = VX$, $dY = V\,dX + X\,dV$

Hence $2\,dX = X\,dV$ and $2\log X = V + A$

Thus $\qquad 2\log(y+1) = \left(\dfrac{2y+x}{y+1}\right) + A$

(c) Re-writing as

$$\frac{dy}{dx} + 2y = x(1+x)$$

we have e^{2x} as an integrating factor and the integral of the left-hand side is ye^{2x}. The right-hand side is

$$\int e^{2x}(x+x^2)\,dx = \frac{(x+x^2)e^{2x}}{2} - \int \frac{e^{2x}}{2}(1+2x)\,dx$$

$$= \frac{(x+x^2)e^{2x}}{2} - \frac{(1+2x)e^{2x}}{4} + \int \frac{e^{2x}}{4}(2)\,dx$$

$$= \left(\frac{2x^2-1}{4}\right)e^{2x} + \frac{e^{2x}}{4} + A = \frac{x^2 e^{2x}}{2} + A$$

Hence the solution is

$$ye^{2x} = \frac{x^2 e^{2x}}{2} + A$$

(d) If

$$\frac{dy}{dx} + 4y = x^3$$

then e^{4x} is an integrating factor and integrating gives

$$ye^{4x} = \int x^3 e^{4x}\,dx = e^{4x}\left[\frac{x^3}{4} - \frac{3x^2}{16} + \frac{3x}{32} - \frac{3}{128}\right] + A$$

EXERCISE 9.5

In each case put $y=e^{mx}$ etc and form the reduced equation am^2+bm+c. The roots are m_1 and m_2

(a) $m_1=4,$ $\quad m_2=-3,$ $\quad y=k_1 e^{4x} + k_2 e^{-3x}$

(b) $m_1=1,$ $\quad m_2=2,$ $\quad y=k_1 e^{2x} + k_2 e^{x}$

(c) $m_1=3,$ $\quad m_2=3,$ $\quad y=(k_1+k_2 x)e^{3x}$

(d) $m_1=1,$ $\quad m_2=1,$ $\quad y=(k_1+k_2 x)e^{x}$

(e) $m_1=4,$ $\quad m_2=-4,$ $\quad y=k_1 e^{4x} + k_2 e^{-4x}$

(f) $m_1=4i,$ $\quad m_2=-4i,$ $\quad y=k_3 \cos 4x + k_4 \sin 4x$
$\qquad\qquad\qquad\qquad\qquad = A \cos(4x-\epsilon)$

(g) $m_1=-1+3i,$ $\quad m_2=-1-3i,$ $\quad y=e^{-x}(k_3 \cos 3x + k_4 \sin 3x)$
$\qquad\qquad\qquad\qquad\qquad\qquad = Ae^{-x} \cos(3x-\epsilon)$

EXERCISES 9.7

1. (a) Complementary function: $m_1 = 8$, $m_2 = 2$

$$y = k_1 e^{8x} + k_2 e^{2x}$$

Particular integral: try $y = K_1 + K_2 x$
then $y = 5/64 + x/8$
Solution is

$$y = k_1 e^{8x} + k_2 e^{2x} + \frac{5}{64} + \frac{x}{8}$$

The initial conditions give $k_1 = 0.1979$, $k_2 = -0.3542$

(b) Complementary function [from Exercise 9.5 (e)],

$$y = k_1 e^{4x} + k_2 e^{-4x}$$

Particular integral: try $y = ae^x$, then $y = -e^x/15$.
Solution is

$$y = -\frac{e^x}{15} + k_1 e^{4x} + k_2 e^{-4x}$$

The initial conditions give $k_1 = \frac{3}{4}$, $k_2 = \frac{1}{4}$.

(c) Complementary function [from Exercise 9.5 (d)],

$$y = (k_1 + k_2 x)e^x$$

Particular integral: try $y = ae^{2x}$; then $y = e^{2x}$
Solution is $y = (k_1 + k_2 x)e^x + e^{2x}$
and the initial conditions give $k_1 = 1$, $k_2 = -1$.

(d) Complementary function: $m_1 = -1 + 2i$, $m_2 = -1 - 2i$:

$$y = Ae^{-x} \cos (2x - \epsilon)$$

Particular integral: try $y = ax + b$; then $y = \dfrac{x}{5} - \dfrac{2}{25}$

Solution is $y = \dfrac{x}{5} - \dfrac{2}{25} + Ae^{-x} \cos (2x - \epsilon)$

The initial conditions give $A \cos (-\epsilon) = 0$
$$1 = -2A \sin (-\epsilon)$$
and hence $\epsilon = -\pi/2$ and $A = -\frac{1}{2}$.

331

(e) Complementary function: $m_1 = 0$, $m_2 = 1$

$$y = A + Be^x$$

Particular integral: try $y = a \cos x + b \sin x$; then

$$y = -\cos x - \sin x.$$

Solution is $y = A + Be^x - \cos x - \sin x$
Initial conditions give $A = -1$, $B = 3$.

2. Solve homogeneous part by putting $y = e^{mx}$ etc; then $m_1 = 2$, $m_2 = -3$ and

$$y = Ae^{2x} + Be^{-3x}$$

Particular integral: try $y = a + bx + cx^2$; then
$a = -7/6$, $b = -1$, $c = -3$.
Solution is $y = Ae^{2x} + Be^{-3x} - (\frac{7}{6}) - x - 3x^2$
Initial conditions give $A = \frac{11}{10}$, $B = \frac{1}{15}$.

EXERCISES   10.7
1. The general solution of $Y_{t+1} = \lambda Y_t + K$ is

$$Y_t = A\lambda^t + \frac{K}{1-\lambda}$$

where A is a constant

(a) $\lambda = 1.5$, $K = 3$, $Y_t = A(1.5^t) - 6$
    Initial condition gives $A = 8$

t	0	1	2	3	4	5
Y_t	2	6	12	21	34.5	54.75

(b) $\lambda = 0.9$, $K = 2$, $Y_t = A(0.9^t) + 20$
    Initial condition gives $A = -17$

t	0	1	2	3	4	5
Y_t	3	4.7	6.23	7.61	8.85	9.96

(c) $\lambda = 1$, $K = 0$, $Y_t = A$.

Initial condition gives $A = 6$

t	0	1	2	3	4	5
Y_t	6	6	6	6	6	6

(d) $\lambda = 1.1$, $K = 4$, $Y_t = A(1.1^t) - 40$

Initial condition gives $A = 41$

t	0	1	2	3	4	5
Y_t	1	5.1	9.61	14.57	20.03	26.03

2. Let Y_t be the value of the savings after t years; then $Y_0 = 15$, $Y_1 = 1.05(15) + 15$ and in general $Y_{t+1} = 1.05\,Y_t + 15$. The solution is $Y_t = A(1.05)^t - 300$ and $Y_0 = 15$ gives $A = 315$. $Y_{20} = 535.8$.

3. The difference equation is $p_t = -1.5p_{t-1} + 3$ and the solution is $p_t = A(-1.5)^t + 1.2$, which is 'exploding' since $1.5 > 1$.

4. The difference equation is $p_t = 0.5p_{t-1} + 3$ with the solution $p_t = -(-0.5)^t + 2$, which is stable since $0.5 < 1$. The equilibrium price is 2.

5. The difference equation is $p_t = -p_{t-1} + 8/3$ with the solution $p_t = [2(-1)^t + 4]/3$ which oscillates since $|-1| = 1$.

6. The difference equation is $0.9\,Y_t = 0.3\,Y_t - 0.3\,Y_{t-1}$ or $Y_t = -0.5\,Y_{t-1}$, and the solution is $Y_t = 100\,(-0.5)^t$.

EXERCISES 10.11

1. Difference equation is $Y_t = (c + b)\,Y_{t-1} - b\,Y_{t-2}$ and putting $Y_t = \lambda^t$ gives the characteristic equation $\lambda^2 = (c + b)\lambda - b$

   (a) Roots are $\lambda_1 = 0.7 + 0.1i$, $\lambda_2 = 0.7 - 0.1i$
   (b) Roots are $\lambda_1 = 0.8$, $\lambda_2 = 0.8$
   (c) Roots are $\lambda_1 = 0.845$, $\lambda_2 = 0.355$.

2. Solve homogeneous part by putting $Y_t = \lambda^t$ and obtaining the roots λ_1 and λ_2 of the characteristic equation. The equilibrium or

particular solution is found by trying expressions of the same form as $f(t)$.

(a) $\lambda_1 = 2$, $\lambda_2 = -1$. Try $Y_t = Z$ as particular solution then $Y_t = -1.5$. Solution is $Y_t = A(2^t) + B(-1)^t - 1.5$ and initial values give $A = 7/3$, $B = 7/6$.

(b) $\lambda_1 = 2$, $\lambda_2 = 1$. Try $Y_t = Z$ as particular solution—fails. Try $Y_t = Zt$; $Y_{t+1} = Zt + Z$, $Y_{t+2} = Zt + 2Z$, then $Z = -4$. Solution is $Y_t = A(2^t) + B - 4t$ and initial conditions give $A = 5$, $B = -4$.

(c) $\lambda_1 = 2$, $\lambda_2 = 2$. Try $Y_t = Z(3^t)$; then $Z = 1$.
Solution is $Y_t = (k_1 + k_2 t)(2^t) + 3^t$ and initial conditions give $k_1 = 2$, $k_2 = -1.5$.

(d) $\lambda_1 = 2$, $\lambda_2 = 1$. Try $Y_t = Zt(2^t)$; then $Z = +1.5$. Solution is $Y_t = A(2^t) + B + 1.5t(2^t)$ and initial conditions give $A = 1$, $B = 0$.

(e) $\lambda_1 = -0.5$, $\lambda_2 = -0.5$. Try $Y_t = Z_0 + Z_1 t$; then $Z_0 = -\frac{2}{9}$, $Z_1 = \frac{1}{3}$. Solution is $Y_t = (k_1 + k_2 t)(-0.5)^t - \frac{2}{9} + (t/3)$ and initial conditions give $k_1 = -88/9$, $k_2 = 8$.

(f) $\lambda_1 = -2 + i$, $\lambda_2 = -2 - i$ hence $a = -2$, $b = 1$, $r = \sqrt{5}$ and $\cos\theta = -2/\sqrt{5}$ so $\theta = -26° \, 34' = 0.4625$ radians. Try $Y_t = Z$; then $Z = 0.4$ and the solution is
$$Y_t = A(\sqrt{5})^t \cos(-0.4625t - \epsilon) + 0.4$$

(g) $\lambda_1 = 2 + 2i$, $\lambda_2 = 2 - 2i$, hence $a = 2$, $b = -2$, $r = \sqrt{8}$ and $\cos\theta = 2/\sqrt{8} = 0.7071$, so $\theta = \pi/4$. Try $Y_t = k_1 + k_2 t$ and $k_1 = 2/25$, $k_2 = \frac{1}{5}$. The solution is $Y_t = A(\sqrt{8})^t \cos[\pi(t/4) - \epsilon] + (2/25) + (t/5)$.

3. Equation is $(1 - \alpha)Y_t = (B + \gamma)Y_{t-1} - \gamma Y_{t-2} + \delta$
   (a) $\lambda_1 = 0.625 + 0.599i$, $\lambda_2 = 0.625 - 0.599i$
   hence $a = 0.625$, $b = 0.599$, $r = 0.8657$, $\theta = 0.7640$ radians. Try $Y_t = Z$ and $Z = 5$.
   Solution is $Y_t = A(0.8657)^t \cos(0.7640t - \epsilon) + 5$.

   (b) $\lambda_1 = 0.4 + 0.49i$, $\lambda_2 = 0.4 - 0.49i$; hence $a = 0.4$, $b = 0.49$, $r = 0.6324$, $\theta = 0.8860$ radians. Try $Y_t = Z$ and $Z = 3.33$. The solution is $Y_t = A(0.6324)^t \cos(0.886t - \epsilon) + 3.33$.

Further Reading

This text has covered the minimum mathematics required by economists. There are many other texts available which cover similar material in differing amounts of detail. Two recommended ones are:

Allen, R. D. G. *Mathematical Analysis for Economists* (Macmillan, 1938)

Lewis, J. Parry. *Introduction to Mathematics for Students of Economics* (Macmillan, 2nd ed 1969)

The second edition of Parry Lewis's text is particularly strong on linear algebra.

Many texts are now available which approach economics from the mathematical viewpoint. For example:

Allen, R. D. G. *Macroeconomic Theory* (Macmillan, 1967)

Baumol, W. J. *Economic Theory and Operations Analysis* (Prentice-Hall, 3rd ed 1972).

Bowers, D. A., and Baird, R. N. *Elementary Mathematical Macroeconomics* (Prentice-Hall, 1971).

One of the main uses of mathematics is in the development of models of economic systems. This is covered in A. R. Bergstrom, *The Construction and Use of Economic Models* (English Universities Press, 1967) where calculus and in particular differential equations are used. Difference equations are used in a similar way in W. J. Baumol, *Economic Dynamics* (Macmillan, 2nd ed 1959), and the role of probabilistic factors is considered in C. F. Christ, *Econometric Models and Methods* (John Wiley, 1966).

The main application of matrix algebra in economics is in the field

335

of econometrics and a recommended text, which includes an explanatory chapter on matrix algebra, is:

Johnston, J. *Econometric Methods* (McGraw-Hill, 2nd ed 1972)

Other applications of mathematics in economics can be found in almost any current economics journal.

Index